Analog Circuits and Signal Processing

For further volumes:
http://www.springer.com/series/7381

Vibhu Sharma · Francky Catthoor
Wim Dehaene

SRAM Design for Wireless Sensor Networks

Energy Efficient and Variability Resilient Techniques

Springer

Vibhu Sharma
ESAT-MICAS
K.U. Leuven
Heverlee
Belgium

Francky Catthoor
Departement ESAT
IMEC
Heverlee
Belgium

Wim Dehaene
ESAT-MICAS
K.U. Leuven
Heverlee
Belgium

ISBN 978-1-4614-4038-3 ISBN 978-1-4614-4039-0 (eBook)
DOI 10.1007/978-1-4614-4039-0
Springer New York Heidelberg Dordrecht London

Library of Congress Control Number: 2012942704

Printed on acid-free paper

Springer is part of Springer Science+Business Media (www.springer.com)

Preface

The text in front of you, dear reader, is the result of 4 years of Ph.D. research. The question we asked ourselves is: Can we drastically reduce the energy consumption of the RAMs in sensor nodes as they are conceived today? We were specifically thinking about body area networks but the research results are not limited to those networks. Any application that lives or has to live, with low speed, in the order of 100 MHz processing will profit from what we have presented here.

The approach we took can brute-force be described as: Let us question everything that designers normally do when setting up static RAM and see what there is to gain. No more givens, no more certainty. Formulated like this, it sounds pretty obvious but this definitely was not the case. Static RAMs are complex, rather analog, circuits. Everything tends to depend on everything. So, the research becomes a story of a very interesting, yet difficult, trade of balancing. A story of carefully analyzing the dynamic versus static energy tradeoff while selecting supply and threshold voltages. A story of trading dynamic energy versus speed performance when setting up local blocks and data transfer architectures. And, this is only the proverbial tip of the iceberg. We chose not to handle these tradeoffs in a theoretical way. This would lead to generic but hard to concretely apply results. The approach was bottom-up. We started from well-defined cases and specifications, analyzed the corresponding tradeoffs, and solved them. At the end of that design research we generalize.

In this book, we will thus explain the different tradeoffs in the design of relatively slow but ultra energy-efficient memories. We hope it will help you in your own quest toward ultra energy-efficient static RAM. It is the result of the designs we performed at KU Leuven in cooperation with the sensor node designers of imec (Holst Centre). The authors wish to thank our imec colleagues for this fruitful cooperation. It was very effective and it made sure that our specifications and the corresponding results are application relevant. We also present silicon proven results. Remember, memories are quite analog circuits so when pushing the limits only one final verification step can give definitive answers: implementation and measurement. That is what we did. You will find the test results of SRAM prototypes in 90 and 65 nm in this book.

Ph.D. research is also a story of individuals. A journey of a young researcher and a less young advisor. A journey along beliefs, doubts, skepticism maybe, hypotheses, simulation, and measurement verification and in the end conclusions. Only if these two individuals get along well in the travel, the story becomes nice to read. From our side, the journey was a worthwhile learning experience. We want to share it with you.

Leuven, Belgium, April 2012

Vibhu Sharma
Francky Catthoor
Wim Dehaene

Contents

Acronyms

BL	Bit Line
RBL	Read Bit Line
WBL	Write Bit Line
GRBL	Global Read Bit Line
GWBL	Global Write Bit Line
WL	Word Line
RWL	Read Word Line
WWL	Write Word Line
LWL	Local Word Line
GWL	Global Word Line
GRWL	Global Read Word Line
GWWL	Global Write Word Line
VGBL	Vertical Global Bit Line
HGBL	Horizontal Global Bit Line
VGWBL	Vertical Global Write Bit Line
HGWBL	Horizontal Global Write Bit Line
VGRBL	Vertical Global Read Bit Line
HGRBL	Horizontal Global Read Bit Line
LBL	Local Bit Line
LWBL	Local Write Bit Line
LRBL	Local Read Bit Line
HVT	High Vt (threshold)
LVT	Low Vt (threshold)
SVT	Standard Vt (threshold)
PDF	Probability Density Function
CDF	Cumulative Distribution Function
DRC	Design Rule Checks
SA	Sense Amplifier
SRAM	Static Random Access Memory
DSM	Deep Sub-micron
GSA	Global Sense Amplifier

SNM	Static Noise Margin
WTP	Write Trip Point
WM	Write Margin
VDD	The power supply voltage
VSS	The negative power supply voltage
Vt	Transistor threshold voltage

Chapter 1
Introduction

1.1 Motivation and Objectives

The development of wireless sensor networks has revolutionized our lifestyles. The sensor networks can be used for various applications like military surveillance, environment monitoring, medical diagnosis etc. The body area network is defined as a wireless sensor network used for medical diagnosis (Istepanian et al. 2004; Gyselinckx et al. 2005). The body area networks have to do continuous health monitoring and to provide real-time feedback. The body area networks facilitate continuous monitoring of the physiological parameters. This continuous monitoring for the large time intervals in the natural environment offers better results compared to the physiological parameters obtained from the short duration monitoring for e.g. stays at a hospital (Park et al. 2003). In order to further extend the capabilities of the body area network, miniature wireless sensor nodes with the extended operational life are required. The sensor nodes have to be of very small form factor (<1 cm^3) (Gyselinckx et al. 2005) for realizing ubiquitous sensing, without interfering with the object being monitored. This miniaturization results in reduced on sensor energy capacity because the size of the battery used to store the energy gets limited. The requirement of invasive surgery (Malan et al. 2004) complicates the battery replacement of the implanted medical wireless sensor nodes.

Energy scavenging from the operating environment can extend the sensor node lifetime for a given battery capacity. Theoretically if the energy scavenged during the operation of the system is larger than the average consumed energy then the sensor nodes could operate forever. But the amount of energy scavenged is also limited (Gyselinckx et al. 2005). Furthermore, power dissipated by these nodes produce heat which is absorbed by the body tissues and increases the temperature of the body (Malan et al 2004). The limited energy source and the heat dissipation limit require the medical wireless sensor nodes to be highly energy efficient. The target energy consumption has to be below 100 uW/cm^2 (Declerck 2005).

V. Sharma et al., *SRAM Design for Wireless Sensor Networks*,
Analog Circuits and Signal Processing, DOI: 10.1007/978-1-4614-4039-0_1,

Table 1.1 Percentage contribution of SRAM in the total dynamic energy consumption for biomedical microprocessors

SRAM in uP (nm)	Capacity	Dynamic energy (%)
(B.Zhai et al. 2006), 130	2 Kb SRAM	47
[NIckes08], 90	60 Kb SRAM	50
(Kwong et al. 2009), 65	128 Kb SRAM	63

Energy efficient sensor networks design involves a holistic approach, covering all aspects of the senor network viz. network protocols, software and hardware platforms. The RF communication consumes a major proportion of the power budget. Even with the most energy efficient transceivers based on ultra-wide-band technology. The power consumption is tens of mill watts (Ryckaert et al. 2007), which is much higher than the available power budget. Moreover, the continuous transmission of a sensed raw data would lead to data congestion in sensor network, thereby increasing the data latency.

This problem is remedied by providing a computational intelligence at the sensor node. The signal processing performed on the sensor avoids the energy expensive raw data transmission. The digital signal processing executed on a platform on a node (master node) compresses the raw data before transmission. Several millions of operations per second performed for processing of the raw data reduce the amount of data transmission. The computation intensive wireless sensor nodes with the increased memory sizes significantly reduce the requirement of data transmission. This trade off between raw data transmission and on senor computation places a higher burden on energy—efficient computation. In order to implement complex digital signal processing algorithms increased amount of memories for the microprocessors are required.

The software code optimization techniques (Verma and Marwedel 2007), tend to improve locality of data/instruction fetches. Hence, in the memory hierarchy system the largest memories have the least number of accesses per word whereas the largest numbers of accesses per word are situated in the L1 memory. With the result L1 embedded memory design becomes a key element for meeting the power budget for the computation intensive wireless sensor nodes.

The on-chip SRAM cache consumes a major proportion of the total dynamic energy per operation (Table 1.1). Zhai et.al. (2006) had proposed a sensor processor with a KHz performance requirement and with a very small capacity 2 Kb SRAM. Even then dynamic energy contribution of SRAM is 47 % of the total dynamic energy consumption of sensor processor operating at VDD = 0.4 V. The static energy consumption is also very important for the wireless sensor nodes, especially because of the long idle periods of operation. Table 1.2 shows percentage contribution of SRAM in the total static energy for the wireless biomedical processors. Even with the biomedical processors designed with older technology nodes like 180 nm, which exhibits lower sub threshold leakage. The total static energy consumption is predominantly coming from SRAM ($\sim$80–90 %).

Table 1.2 Percentage contribution of SRAM in the total static energy consumption for bio-medical microprocessors

SRAM in uP (nm)	Capacity	Static Energy (%)
(M.Seok et al. 2008), 180	52 × 40 bit DMEM	89
	64 × 10 bit IMEM	
(G.Chen et.al. 2010), 180	24 Kb SRAM	79

With technology scaling wireless sensor node form factor is not limited by the size of the integrated circuits. The total size of the wireless sensor node is limited by the energy carrying capacity (battery sizes). The tradeoff of integrated circuit (SRAM macro) area for achieving lower energy consumption is a viable option for reducing the overall form factor of the wireless sensor nodes. The increased energy efficiency of the integrated circuits (SRAM macro) would require smaller sized batteries.

The performance requirement for the wireless sensor nodes are bit relaxed because processing of the low speed signals is required. The operational clock frequency of the sensor nodes is dependent on the application scenario ranging from a few hundreds of KHz (Zhai et.al. 2006; Chen et.al. 2010) to few MHz (Kwong et al. 2009) and the tens of MHz range (Ickes et.al. 2011; Hulzink et al. 2011). Therefore the performance requirements are not very stringent compared to the high performance multimedia applications.

The energy efficient SRAM design is a key requirement to enable further extensions of the capabilities of the energy limited wireless sensor nodes. Therefore, ultra low energy SRAMs which prioritize energy efficiency over performance and area overhead are required for body area networks.

1.2 Traditional SRAM Design and Technology Scaling

SRAM design requires complex trade off between process induced limitations (process variations) and system level design limitations (area overhead and energy consumption). Technology scaling reduced the energy consumption of both memories and processor on a regular basis. Energy reduced in proportion to λ^3, with λ the smallest feature size that can be realized in the technology. The memories are the most vulnerable to ever increasing process variations in advanced technology nodes. This ended with the introduction of the 90 nm technology node (Gielen and Dehaene 2005).

Figure 1.1, illustrates low energy SRAM design challenges. SRAM bit cell functional parameter degradation due to increasing variability and voltage scaling is of utmost concern. The process variations are classified into two categories one which result in the differences in the characteristics of the neighboring devices on the same die (intra die) and the other which effects all the devices on a die in the same manner (inter die). The intra die variations are in inverse proportion to the

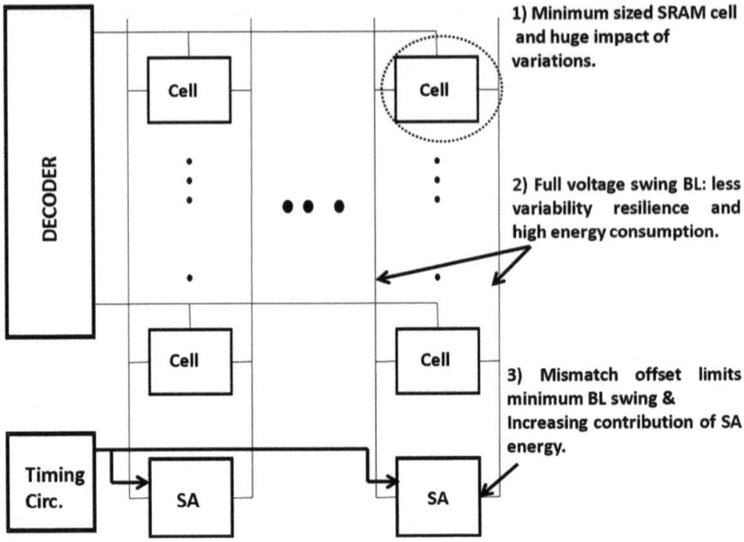

Fig. 1.1 Low energy variability resilient sram design challenges

square root of the transistor channel area (Shyu et al. 1982; Pelgrom et.al. 1989). The shrinking transistor dimensions with the technology scaling result in a large intra die variations (Asenov 1998). The increasing intra die variations degrade I_{READ}, SNM_{READ} and the write margin of the SRAM cell (Yamaoka et.al. 2004; Yamaoka et al. 2006).

The intra die variations also impact the performance of sense amplifier circuits (Zhang et.al. 2000). The intra die variations are the main source of the SA offset voltage, which puts a lower bound on the input signal. In other words, the ability of a sense amplifier to sense a small bit-line swing is limited by the distribution of its offset voltage. At the same time the bit-line discharge must be minimized not only to maintain the performance but also to reduce the energy consumption associated with the charging and discharging of highly capacitive bit-lines. Alternatively, enabling low swing sensing by using upsized sense amplifier critical transistors in order to reduce the mismatch offsets (Pelgrom et.al. 1989) increases the sense amplifier energy consumption. This is problematic, in particular for the advanced technology nodes where the contribution of the sense amplifier energy to the total READ energy is becoming more prominent (Cosemans et al. 2009).

The classic design paradigms in order to meet the challenges posed by the technology scaling rely on the upsizing and on the extra design margins. This over sizing and too much insertion of the design margins result in an excessive degradation of the energy consumption and performance. For example, word line signal pulse width is dictated by the worst cell in the memory word. But in the time duration dictated by the worst SRAM cell, the average SRAM cells have already discharged the bit-lines unnecessarily more than the requirement. Figure 1.2, illustrates distribution of bit-line swing for the word line pulse width, decided by

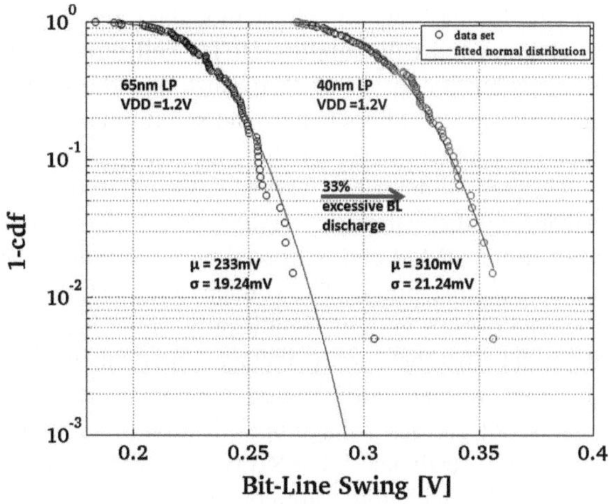

Fig. 1.2 Distribution of bit-line voltage swing for the column height of 512 cells, VDD = 1.2 V, Nominal process corner. SRAM cell assertion pulse width is adjusted to ensure target 150 mV bit-line discharge for the slowest 6T SRAM cell

the worst case SRAM cell to achieve 150 mV of required bit-line discharge. Therefore, the circuit design techniques which improve the operating margins of SRAM without increasing the energy consumption are required.

The circuit techniques include: innovation in the local architecture with the use of local RD/WR assist circuitry for the conventional 6T SRAM cell. The energy efficient hierarchical bit-lines structure includes low swing global bit-lines and VDD/2 pre-charged short local bit-lines. The proposed reduced swing dual VT 8T SRAM cell solves the issues related with the conventional 6T SRAM cell. At the same time it also reduces the leakage associated with the 8T SRAM cell. The innovative Multi-Sized SA redundancy calibration technique for the SA yields maximum energy reduction compared with the existing calibration techniques. The novel Charge Limited Sequential SA solves energy-offset issue without relying on the calibration phase. These techniques have been implemented and the prototypes developed are discussed in Chap. 7.

1.3 Structure of the Text

This text discusses vital circuit level techniques for enabling low energy variability resilient SRAM memories for computation intensive wireless sensor nodes.

Chapter 2 describes different SRAM bit cell topologies. First it provides an overview of conventional SRAM 6T cell and its limitation in offering variability resilient operation. Then different SRAM cell topologies are discussed which offers better stability margins compared to SRAM 6T cell. Different cell topologies discussed are broadly categorized as 7T, 8T and 10T. It is extremely difficult for a single cell topology to address all the issues like SNM read, write margin and half select condition. RD8T SRAM cell is very stable but it does not offer much improvement in write-ability and also adds to the leakage consumption. D2APT

8T cell does not rely on voltage modulation for higher write margins but it does not offer improved cell stability. DETG cell offers very high write and cell stability margins but it consists of 10 transistors and also the read sensing is single ended. Z8T SRAM cell offers better read stability and differential sensing but there is an inherent problem in its topology as discussed which limits it applicability.

Chapter 3 describes various dynamic voltage optimization techniques for enhancing the variability resilience of SRAM 6T cell. The minimum supply voltage for SRAM cell is limited by write failures (write—ability) or read disturb failures (cell stability). The voltage optimization can impact the write failures and the read disturb failures significantly as the SRAM 6T cell functionality is highly dependent on the supply voltage. First the implementation details of different dynamic voltage optimization techniques are provided. Then a Comparative analysis of various voltage optimization based assist techniques in improving the variability resilience of SRAM 6T cell is discussed. The adaptive voltage optimization techniques are compared based on the functional effectiveness, performance and the energy consumption. It is observed that the voltage optimization which increases the strength of the NMOS access transistors of the SRAM cell are better compared to the techniques which target reducing the strength of the latch transistors. Alternatively for improving the SNM_{READ} (cell stability). The techniques which increase the strength of the latch transistors are more effective compared to the techniques which reduce the strength of the NMOS access transistors. Finally an overview on new kind of hybrid voltage optimization techniques like configurable write assist, crosshair SRAM, mimicked negative bit-line (MNBL) technique and compounded differential VSS (CDVSS) bias technique is provided. These hybrid techniques combines existing two or more adaptive voltage optimization techniques to yield better performance and solves some of the key issues related to the existing adaptive voltage optimization techniques.

Chapter 4 discusses various logic circuit based assist techniques to alleviate complex design trade off effort of SRAM cell design. The local assist circuit techniques proposed here solves the issues associated with the increased device variations at the scaled voltage levels for the advance sub-nanometre technologies. The different assist techniques discussed are: hierarchical divided bit-lines which reduce the effective bit-line capacitance for accessed SRAM cell and enhance the cell stability. Hierarchical divided bit-lines with local assist circuitry as proposed by adds an upsized low VT read buffer as a local assist circuit in order to accelerate the global bit-line discharge rate and achieves high performance. WRITE after READ based assist circuitry enables DRAM type sensing operation by rewriting the cell content after every READ operation and offers very high cell stability at the expense of increased energy consumption. Low swing local bit-line hierarchy with or without parasitic isolation offers very high SNM read and also reduces the energy consumption. High bit density based bit-line hierarchy also reduces the area overhead.

Chapter 5 discusses the circuit technique and voltage optimization techniques for realizing ultra low energy and variability resilient SRAM's for L1 data and instruction memories. SRAM bit cell functional parameter degradation due to increasing variability and decreasing power supply is of utmost concern. The classic design paradigms in order to meet the challenges posed by the technology scaling rely on the upsizing and on the extra design margins. This over sizing and too much insertion of the design margins result in an excessive degradation of the energy consumption and performance. First the static leakage energy minimization techniques are discussed along with their implementation overhead. Then various WRITE energy reduction techniques are described. The selective bit-line voltage scaling (low swing bit-lines) helps in reducing the energy consumption. The energy consumption is further optimized by having a masking feature. Litho optimized low swing static write signal based techniques reduce the timing complexity and transistor sizes. With the result the energy reduction gain achieved is of the highest order compared to the other state of the art techniques. The chapter concludes with circuit techniques for achieving low energy READ operation.

Chapter 6 addresses the design of low energy variability resilient READ sense amplifier of the memory. It discusses the fundamental limitation on the SA performance, especially for the memories in deep sub micron technologies. It covers various calibration based sense amplifier design techniques for solving energy-offset trade off issue. Then the implementation details for Multi-sized SA redundancy and its comparison with the various calibration based techniques is provided. Then a charge limited sequential sensing concept is discussed for solving energy-offset trade off issue without resorting to calibration. Finally the design, implementation details and the operation of a calibration free sense amplifier based on the charge limited sequential sensing is provided.

Chapter 7 describes two prototypes of SRAM macro which have been successfully developed, fabricated and tested in order to validate the proposed low energy and variability resilient circuit techniques discussed in the previous chapters. First a design overview of IM_90 (first prototype) is provided followed by IM_65 (second prototype). This chapter concludes with the performance comparison of IM_90 and IM_65 with the current state of the art for the wireless sensor node applications.

References

A. Asenov, Random dopant induced threshold voltage lowering and fluctuations in sub-0.1 m MOSFETs: a 3-D atomistic simulation study. IEEE Trans. Electron Devices **45**(12), 2505–2513 (1998)

G. Chen et al., Millimeter-Scale nearly perpetual sensor system with stacked battery and solar cells, in *Proceedings of IEEE International Solid State Circuits Conference (ISSCC)*, pp. 288–289, (2010)

S. Cosemans, W. Dehaene, F. Catthoor, A 3.6pJ/Access 480 MHz, 128 kbit On-Chip SRAM with 850 MHz boost mode in 90 nm CMOS with tunable sense amplifiers, *IEEE J.Solid State Circuits*, pp. 2065–2077, (2009)

G. Declerck, A look into the future of nanoelectronics, *Digest of Technical Papers of the 2005 Symposium on VLSI Technology*, pp. 6–10 (2005)

G. Gielen, W. Dehaene, Analog and digital circuit design in 65 nm CMOS: end of the road? in *Proceedings of Design, Automation and Test in Europe (DATE)*, pp. 37–42, (2005)

B. Gyselinckx et al., Human ++: autonomous wireless sensors for body area networks, in *Proceedings of IEEE Custom Integrated Circuits Conference (CICC)*, pp. 13–19, (2005)

J. Hulzink et al., An ultra low energy biomedical signal processing system operating at near-threshold. IEEE Tran. Biomed. Circuits Syst. **5**(6), 546–554 (2011)

N. Ickes, D. Finchelstein, A.P. Chandrakasan, A 10-pJ/instruction, 4-MIPS Micropower DSP for Sensor Applications, *IEEE Asian Solid-State Circuits Conference*, pp. 289–292, (2008)

N. Ickes et al., A 10pJ/cycle ultra low voltage 32-bit microprocessor system-on-chip, in *Proceedings of IEEE European Solid State Circuits Conference (ESSCIRC)*, pp. 159–162, (2011)

R.S.H. Istepanian et al., Guest editorial introduction to the special introduction to the special section on m-health: beyond seamless mobility and global wireless health—care connectivity. IEEE Trans. Inf Technol. Biomed. **8**(4), 405–411 (2004)

J. Kwong et al., A 65 nm sub-vt microcontroller with integrated SRAM and switched capacitor DC-DC converter. IEEE J. Solid-State Circuits **44**(1), 115–126 (2009)

D. Malan et al., Code blue: an ad hoc sensor network infrastructure for emergency medical care, in *Proceedings of International Workshop Wearable Implantable Body Sensor Network* (2004)

S. Park et al., Enhancing the quality of life through wearable technology. IEEE Eng. Med. Biol. Mag. **22**(3), 41–48 (2003)

M. Pelgrom et al., Matching properties of MOS transistors, *IEEE J. Solid-State Circuits*, pp. 1433–1439 (1989)

J. Ryckaert et al., A CMOS ultra-wide-band receiver for low data-rate communication. IEEE J. Solid-State Circuits **42**(11), 2515–2527 (2007)

M. Seok et al., The Phoenix processor: a 30 pW platform for sensor applications, in *Symposium on VLSI Circuits Digest of Technical Papers*, pp. 188–189, (2008)

J.B. Shyu, G.C. Temes, K. Yao, Random errors in MOS capacitors. IEEE J. Solid-State Circuits **SC-17**(6), 1070–1076 (1982)

M. Verma, P. Marwedel, Advance memory optimization techniques for low-power embedded processors, ISBN 978-1-4020-5896-7, (Springer, Netherlands, 2007)

M. Yamaoka et al., Low power SRAM menu for SOC application using Yin-Yang-feedback memory cell technology, in *Digest of Technical Papers of Symposium on VLSI Technology*, pp. 288–291, (2004)

M. Yamaoka et al., 90 nm process-variation adaptive embedded SRAM modules with power-line-floating write technique. IEEE J. Solid-State Circuits **41**(3), 705–711 (2006)

K. Zhang et al., The scaling of data sensing schemes for high speed cache design in sub-0.18 um technologies, in *Symposium on VLSI Circuits Digest of Technical Papers*, pp. 226–227, (2000)

B. Zhai et al., A 2.60pJ/Inst sub-threshold sensor processor for optimal energy efficiency, in *Symposium on VLSI Circuits Digest of Technical Papers*, pp. 154–155 (2006)

Chapter 2
SRAM Bit Cell Optimization

This chapter discusses different SRAM bit cell topologies. This chapter first provides an overview of the conventional SRAM 6T cell and its limitations. Then different SRAM cell topologies are discussed which offers better stability margins compared to 6T SRAM cell. Different cell topologies discussed are broadly categorized as 7T, 8T, and 10T. It also classifies SRAM cells based on single ended and differential sensing. Finally, the chapter concludes with a summary of different SRAM cells topologies.

2.1 Introduction

The usage of SRAM is continuously increasing in system-on-chip (SOC) designs. Process technology scaling has contributed remarkably in improving the performance of and area density of SOC. The SRAM cell typically utilizes the minimum sized transistor in order to realize a high density. With the result impact of increased intra die variations with the technology scaling is more pronounced on the SRAM cells. With the result SRAM scaling has become extremely difficult in the advanced technology nodes (e.g., 65, 40, or 32 nm LP CMOS technology).

The lowest operational VDD (VDDmin) for embedded memories (SRAM) is limited by either SNMread (cell stability) or write ability [write margin (WM)]. SRAM bit cell functional parameter degradation due to increasing variability and decreasing power supply is of utmost concern. The random threshold variations in subnanometer technologies have resulted in serious yield issues for realizing low VDD READ/WRITE operations with a 6T SRAM cell. Figure 2.1 shows 6T SRAM cell diagram. It relies on rationed operation to achieve the required functionality. The area of an SRAM cell is very important because the cell area contributes significantly to the silicon area. For instance, SRAM L1 caches occupy a significant portion of many designs. The minimum sized 6T cell in 65 nm

V. Sharma et al., *SRAM Design for Wireless Sensor Networks*,
Analog Circuits and Signal Processing, DOI: 10.1007/978-1-4614-4039-0_2,

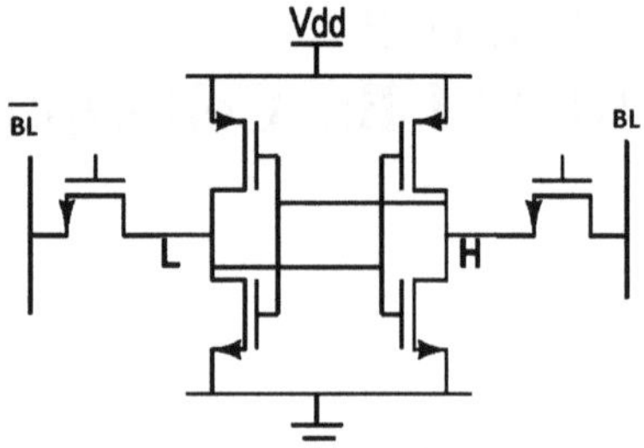

Fig. 2.1 6T SRAM cell

occupies 0.4 μm^2 (Utsumi et al. 2005), in the 40 nm 0.33 μm^2 (Yabuuchi et al. 2007), and in the 32 nm 0.124 μm^2 (Chang et al. 2005).

Impact of Process Variations

As the SRAM cell is scaled, it is difficult to ensure cell stability. For low VDD values, the read SNM becomes negative (loss of bistability). This is because of the reduced signal levels at the low VDD levels and also because of the impact of Vt variations. SRAM cell design can be optimized to minimize the impact of Vt variation on SNMread. The SRAM cell beta ratio is defined as the (W/L) of NMOS pull down transistors of inverter to the (W/L) of nMOS pass transistors. The cell beta ratio balances performance and stability. For stability, increasing the beta reduces the risk of data flip during the READ operation. However, for performance stronger pass transistor is desired. The conventional 6T SRAM cell topology has an inherent disadvantage that it requires a very complex tradeoff between stability (SNMread) and performance (Iread). The higher value of beta favors cell stability but has a negative impact on Iread. Similarly, lower value of beta increases Iread but also increases the risk of data flips (less stable).

There is another problem of write ability, causing write failures in the SRAM cell. A failure to write occurs when the pass transistor is not strong enough to overpower the pull-up PMOS and pull the internal node to ground (writing "0"). The increased strength of pull up PMOS transistors or the decreased strength of NMOS pass transistors due to the process variations impedes the discharge process through the pass transistor. Furthermore, process variations also reduce the trip point of inverter holding the state "H", resulting in the write failure. The current ratio between the pull up PMOS transistors and the pass access NMOS transistors determines the WM. The successful WRITE operation is achieved by increasing the strength of the write access NMOS pass transistors or by decreasing the strength of the pull up PMOS transistors.

Figure 2.2 shows SNMread versus WM for a 6T cell under different PVT conditions. Utilizing high Vt transistors for SRAM cells increase the cell stability but has an adverse impact on the WM. Similarly, low Vt transistors for SRAM cells improve the WM but results in lower stability. Utilizing a high Vt cell decreases the WM by 14 % and low Vt transistors result in the 36 % degradation of cell

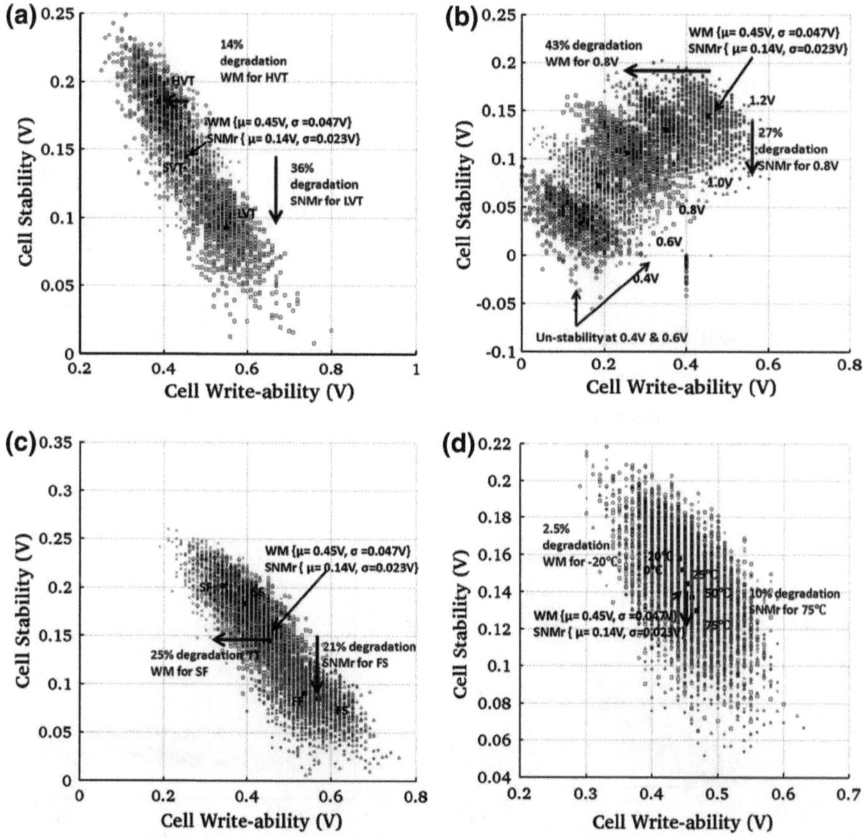

Fig. 2.2 6T cell, 65 nm LP technology, minimum sized SRAM cell. **a** Vt variation. **b** VDD variation for standard Vt 6T cell. **c** Process variation for standard Vt 6T cell. **d** Temp variation for standard Vt 6T cell

stability. Scaling VDD has an adverse impact on both cell stability and WM as explained above because of the increased impact of Vt variations. Regarding inter die variations (considering process corners), slow NMOS (weak pass transistor) and fast PMOS (strong pull up transistor) is the most difficult situation for write ability. This result in 25 % degradation in the WM compared to the nominal process corner. Similarly, from SNMread perspective fast NMOS and slow PMOS results in 21 % degradation in the cell stability. Increasing temperature reduces the Vt of NMOS transistors thereby resulting in reduced cell stability (NMOS pass transistor and NMOS pull down low Vt scenario) by 10 % compared to the nominal temperature.

Similarly, reducing the temperature increases the Vt for NMOS transistors (weak NMOS access transistor) and it results in 2.5 % degradation in WM.

Figure 2.3 shows Iread versus leakage for 6T cell under different PVT conditions. Reducing VDD results in 2.67 orders of magnitude reduction in Iread

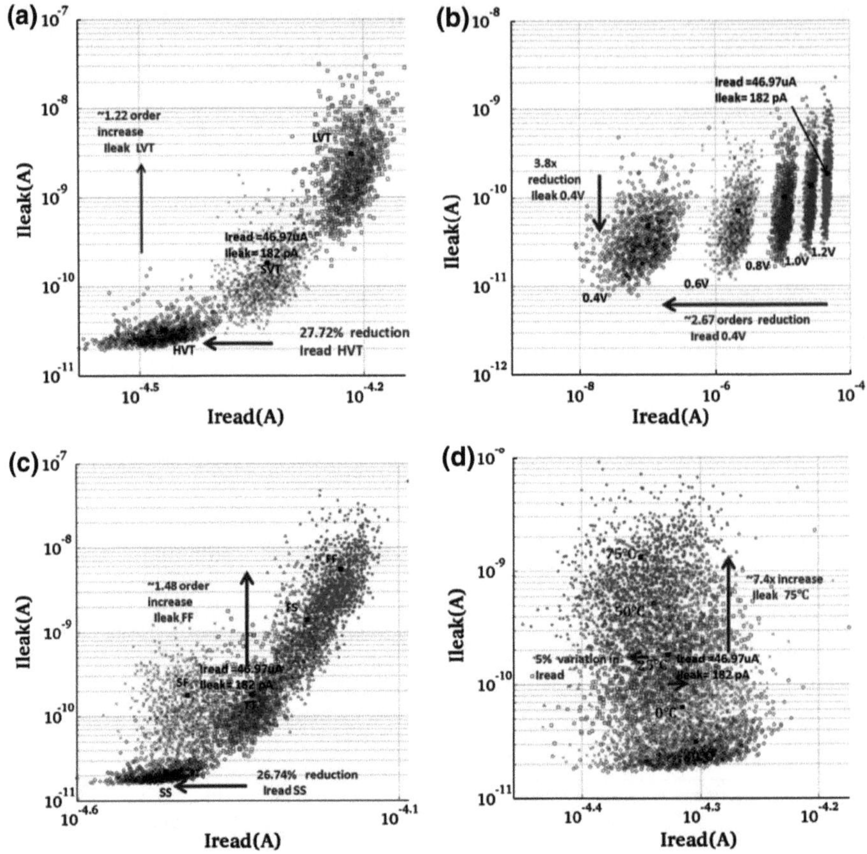

Fig. 2.3 6T cell, 65 nm LP technology, minimum sized SRAM cell. **a** Vt variation. **b** VDD variation for standard Vt 6T cell. **c** Process variation for standard Vt 6T cell. **d** Temp variation for standard Vt 6T cell

at 0.4 V compared to 1.2 V for SRAM 6T cell at 65 nm LP technology node. This serious degradation of Iread at low VDD levels makes SRAM 6T cell less attractive for low VDD applications. The combination of variation on top of dramatically reduced mean Iread means that the read access time is very high, thereby making it less suitable for low VDD applications.

2.2 Different Cell Topologies

This section describes different SRAM cell topologies which solves the issues like degraded SNMread, Iread, WM with 6T SRAM cell for realizing low VDD SRAM.

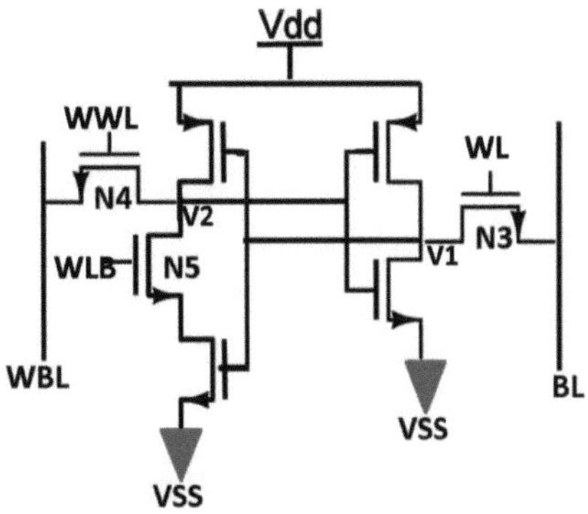

Fig. 2.4 SNMr free 7T cell

2.2.1 Read SNM Free (RSNF) 7T Cell

A transistor N5 is inserted into the 6T cell structure for loop cutting (Fig. 2.4) (Takeda et al. 2006). It enables differential WRITE operation and single ended READ operation. During an idle state when the cell is not accessed WLB is high and the data retention process is exactly same as that of 6T cell. During READ operation WLB is deactivated, the logical threshold voltage of the CMOS inverter driving Node V2 becomes very high (Takeda et al. 2006). Therefore, the SNMread value at V1 = "0" becomes large no matter with the N3 pass transistor activate and increased voltage at node V1. It is difficult to quantify the SNMread with static analysis methods. There is no information provided on how the authors obtained butterfly curves for the read operation. The test chip (Takeda et al. 2006) of 64 Kb RSNF 7T cell macro in 90 nm obtains VDDmin of 0.44 V and 20 ns access time at 0.5 V.

2.2.2 Differential Data Aware Power-Supplied (D^{2}AP) 8T SRAM Cell: Improved Write Margin and Half-Select Accesses

Figure 2.5 shows the D^{2}AP-8T SRAM cell (Chang et al. 2009a). The basic structure is similar to the 6T SRAM cell, except it is powered by its bit-line pair (PSWL and PSWR). During the hold mode the bitlines are kept VDD precharged. The PMOS switches (PSWL and PSWR) are kept ON (ZWL = 0) to power the PUL and PUR of the cross-coupled inverters (invL and invR) from the bit-line pair. The application of differential data-aware (powered by bitlines) voltages to the cross-coupled inverters improve the WM and enlarge the stability margins for half select accesses.

During WRITE operation, the BL is pulled to VSS (writing "0") and BL bar is kept at VDD. The header PMOS switches PSWL and PSWR are ON, the source of invL is reduced (BL pulled to VSS). The trip point of invL becomes lower because of the reduced strength of PMOS transistor of invL. The source voltage of PMOS transistor of invR is at VDD enabling a faster pull up for the complementary node.

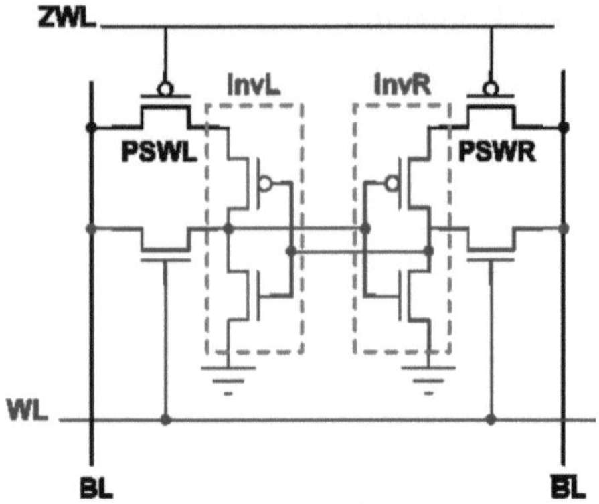

Fig. 2.5 D^2AP-8T cell

The negative feedback mechanism increases the stability margin for half select accesses. Regarding half select condition immunity of the inactivated cells on the asserted word line. If the storage node Q rises being connected with BL and QB is dropped. Then automatically due to the lowering of the BL, it becomes difficult for invL to flip. The READ operation of this cell is similar to the 6T SRAM cell. This cell relies on the boosted bit-line voltage (discussed in Chap. 3) for increasing read cell current and the read static noise margin, especially at lower voltages.

The test chip (Chang et al. 2009a) of 39 Kb SRAM macro featuring D^2AP-8T SRAM cell is fabricated in 40 nm LP CMOS technology. The measured VDDmin of the D^2AP-8T macro is 540 mV. Figure 2.6 shows WM versus VDD for D^2AP-8T and 6T cell at nominal process corner. There are number of issues with D^2AP-8T SRAM cell. The PMOS switches (PSWL and PSWR) of the unselected D^2AP-8T cells on the accessed column are temporarily turned off to isolate the storage nodes from BL during the short BL switching period which increases the risk of data retention. Secondly, the bitlines are kept precharged to VDD (required for powering the inverters of the SRAM cell), increases the stand by leakage power consumption.

2.2.3 Half Select Condition Free Cross Point 8T (CR8T) SRAM Cell

The cross point 8T-SRAM provides two additional access transistor compared to the conventional 6T SRAM cell. It has four access transistors for the Y-address controls as well as the X-address (Fig. 2.7). These access transistors are controlled by the horizontal word line (WLH) and the vertical word line (WLV) signals. For the accessed SRAM cells, both WLH and WLV signals are activated and the internal storage nodes are exposed to the bitlines. For the un-accessed SRAM cells either on the activated column or on the activated row only the WLV is activated or the WLH is activated with the result that internal storage nodes are never exposed to the bit-line information. This is how the half select condition is

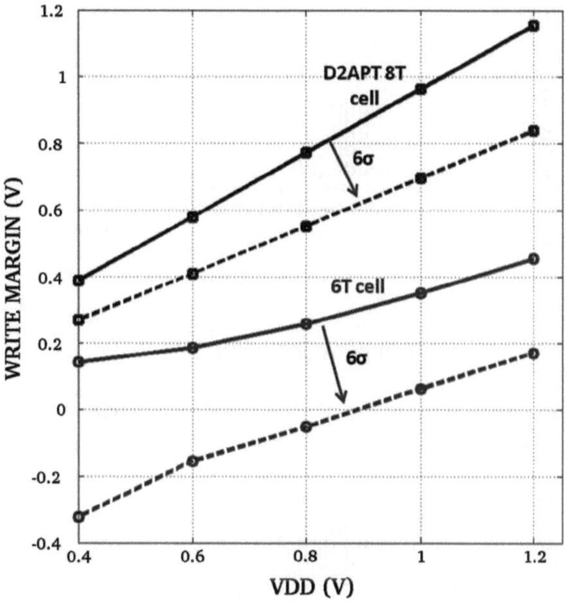

Fig. 2.6 WM versus VDD for 6T SRAM cell and D^2APT 8T SRAM cell in 65 nm LP technology node, nominal process corner, and 25 °C. WM is negative below VDD of 0.9 V for SRAM 6T cell, whereas D^2APT results in positive WM even for 0.4 V

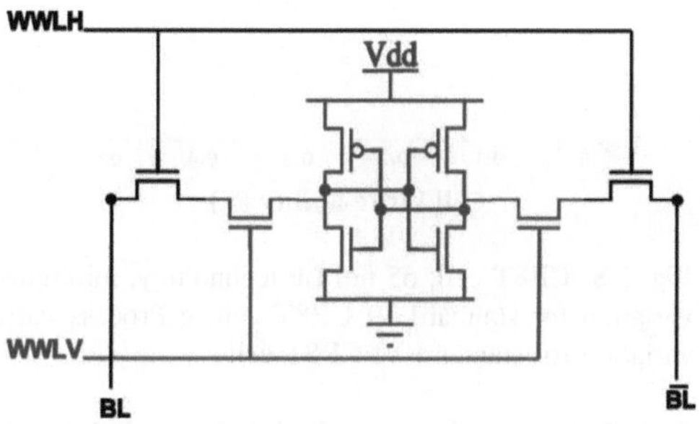

Fig. 2.7 CR8T cell

eliminated. The presence of two NMOS series access transistor also improves the SNMread; it results in 64.3 % improvement in SNMread compared to the same sized 6T SRAM cell. However, two series access transistor results in 44.44 % degradation in the WM and 29.87 % degradation in the cell read current compared to the 6T SRAM cell (Yabuuchi et al. 2009).

The degradation in the cell read current and the WM is addressed by using voltage optimization techniques discussed in Chap. 3. The test chip featuring 1 Mb CR8T SRAM cell along with the negative VSS and the negative bit-line technique achieves VDDmin of 0.6 V in 45 nm LP technology. The negative VSS technique used for the read operation improves read access time by 8.61 ns at 0.6 V and the negative BL technique proposed improves writeability. Figure 2.8 shows SNM-read versus WM for different PVT conditions.

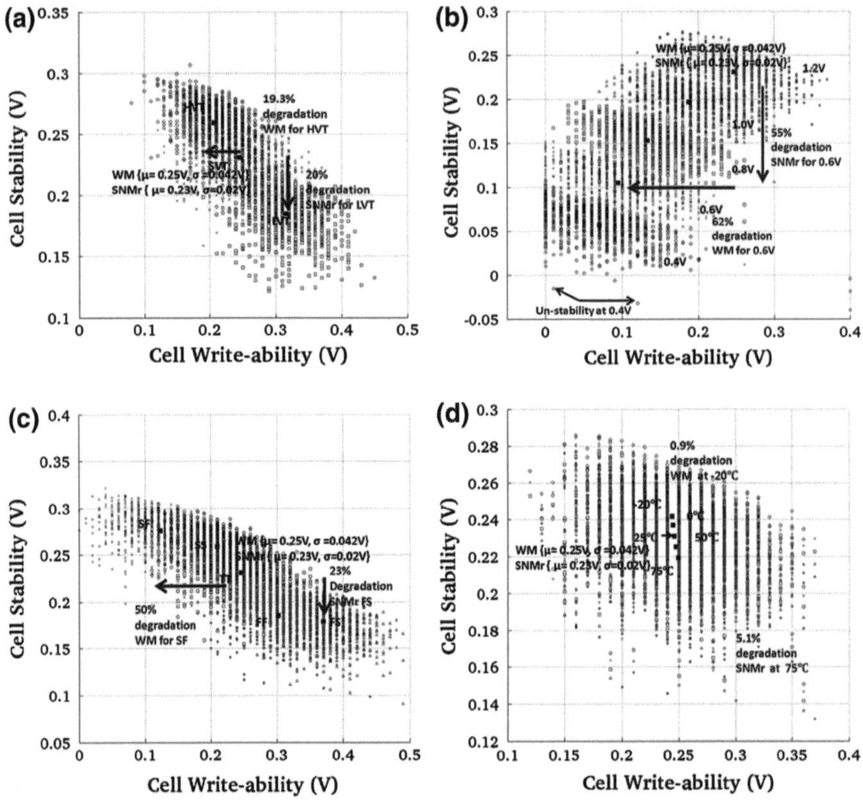

Fig. 2.8 CR8T cell, 65 nm LP technology, minimum sized SRAM cell. **a** Vt variation. **b** VDD variation for standard Vt CR8T cell. **c** Process variation for standard Vt CR8T cell. **d** Temp variation for standard Vt CR8T cell

2.2.4 Read Decoupled 8T and 10T Cell (Isolation of the Internal Storage Nodes from the Read Bit-Lines)

The worst case SNMread in the conventional 6T SRAM cell becomes extremely small with the reduction of the supply voltage refers to Fig. 2.2. With the result 6T SRAM cell cannot be used for the low voltage operations as discussed earlier. This section will discuss different SRAM cells which decouples the cell node from the read bit line by using additional read port transistors. This isolation results in a SNMread equal to the SNMhold (Fig. 2.9).

1. Read decoupled (RD) 8T SRAM cell (Chang et al. 2008)

Figure 2.10 shows RD 8T cell. The structure is similar to the 6T cell except that the two transistors (read stack) are added to a conventional 6T cell. There are separate write and read ports. The word line of the 6T structure is used only during the WRITE operation. Similarly, the word line of the read stack transistors is used

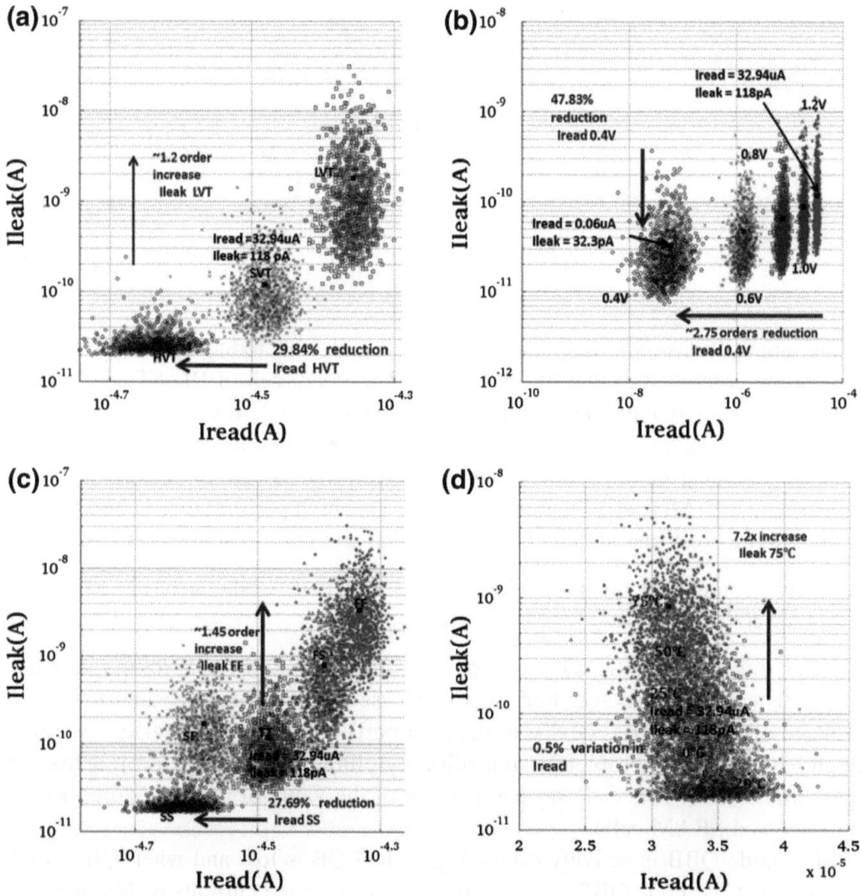

Fig. 2.9 CR8T cell, 65 nm LP technology, minimum sized SRAM cell. **a** Vt variation. **b** VDD variation for standard Vt CR8T cell. **c** Process variation for standard Vt CR8T cell. **d** Temp variation for standard Vt CR8T cell

only for the READ operation. The read word line (RWL) runs parallel to the write word line (WWL). The decoupled read ports eliminate bit-line charge sharing with SRAM internal storage nodes. It avoids read disturb issue for the activated word line. In other words, hold SNM of cell is same as the SNM read. The area overhead of RD-8T SRAM cell is 30 % compared to the 6T SRAM cell (Chang et al. 2008).

The worst case SNMread of RD 8T cell is $\sim 2.1\times$ compared to the SRAM 6T cell SNMread at VDD = 1.2 V. Further, the WM can be improved by increasing the strength of the pass transistors of the write port. The read performance Iread, cell is determined by the strength of the read stack transistors. In this analysis the transistor sizes are kept same (minimum sized); therefore, the WM values are in the same order as that of the SRAM 6T cell. Figure 2.11 shows SNMread versus WM for different PVT conditions. The RD 8T SRAM cell is more suitable for low

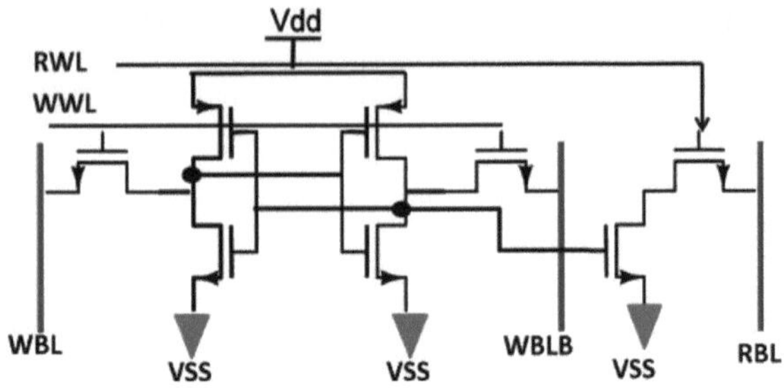

Fig. 2.10 RD 8T cell

VDD applications. The degradation in Iread cell with reducing voltage level is much less compared to that of the 6T cell. The Iread, cell of RD 8T cell is 10.98 and 18.19 uA at 0.4 and 0.6 V compared to the 0.1 and 2.18 uA for the SRAM 6T cell. Figure 2.12 and Iread, cell versus leakage for different PVT conditions. The test chip (Chang et al. 2008) macro of 32 Kb RD 8T SRAM macro in 65 nm operates at 295 MHz at VDDmin of 0.41 V. This high performance is also because of the divided bit-line architecture used in the test chip (discussed in Chap. 4).

2. Data independent bit-line leakage (DIL) 10T cell (Calhoun et al. 2006; Kim et al. 2007)

The single ended READ operation with RD 8T SRAM cell results in a data dependent bit-line leakage. For the worst case scenario (stored value Q = "L", voltage drop across pass device of the read port) the increase in leakage can be as high as 30 %. This problem is remedied by eliminating voltage drop across the pass transistor of the read port irrespective of the value of the data stored for the un-accessed SRAM cells.

The node QBB is actively driven high when QB is low and when QB is high, the value at the node QBB is set by the relative leakage currents of M9 and M10 (Fig. 2.13). The threshold voltage of M9 is taken to be lower compared to the NMOS devices M10 and M7. With the result leakage current of M9 is higher compared to the NMOS M10 device and the node QBB approach to VDD. This is how the voltage drop across M8 pass transistor of the read port for the un-accessed SRAM cells remains zero irrespective of the stored value QB. However, this structure is less robust for the skewed process corners where the PMOS strength is less compared to the NMOS strength.

The DIL 10T cell (Kim et al. 2007) (Fig. 2.14) provides a more variability resilient solution. The node A (QBB) is actively driven high independent of the stored data value by turning ON PMOS transistor M10 for un-accessed SRAM cells. The DIL 10T cell results in 55.5× reduction in the bit-line leakage for the same value of the cell read current. The test chip of 480 kb DIL 10T cells in 130 nm technology achieves VDDmin of 0.2 V operating at 120 kHz. The test chip (Calhoun et al. 2006) of 256 kb 10T cells in 65 nm technology achieves VDDmin of 0.4 V operating at 475 kHz. However, the area overhead with read decoupled 10T cells is extremely high (Fig. 2.15).

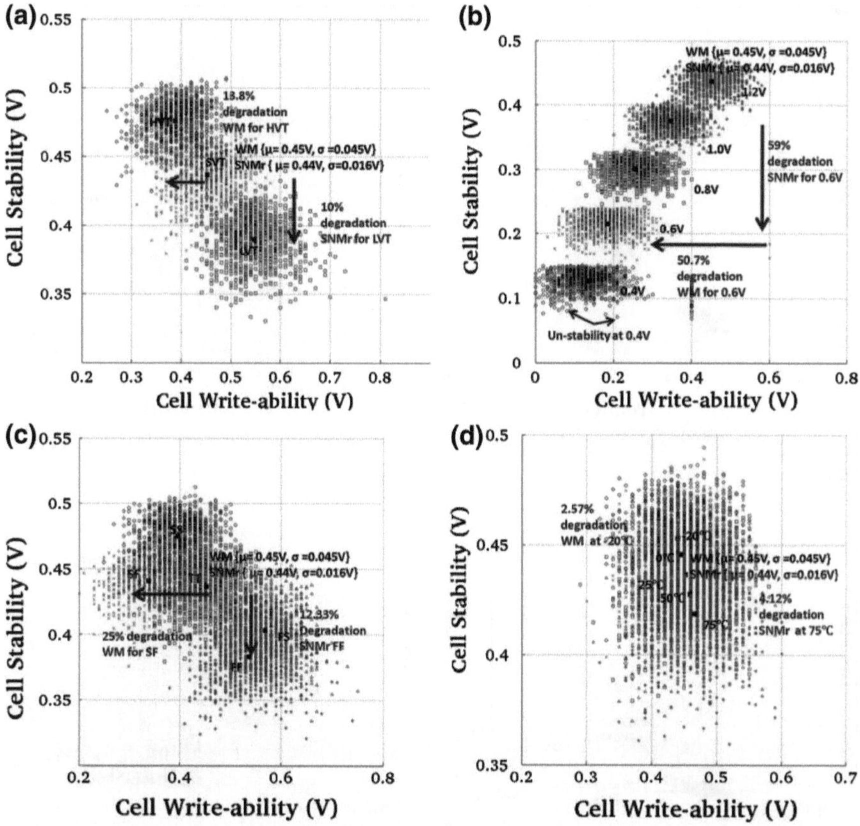

Fig. 2.11 RD8T cell, 65 nm LP technology, minimum sized SRAM cell. **a** Vt variation. **b** VDD variation for standard Vt RD8T cell. **c** Process variation for standard Vt RD8T cell. **d** Temp variation for standard Vt RD8T cell

3. Reduced Swing Dual Vt (RSDVt) 8T SRAM Cell (Sharma et al. 2011b)

Traditional 8T SRAM cells decouple the read port from the internal nodes, thereby eliminating the risk of instability during the read operation. The cell has separate read and WWLs, as well as separate read and write bitlines. The read port consists of two stacked NMOS transistors which deliver the cell read current (I_{READ}) when the RWL is asserted. The two stacked NMOS transistors introduce an additional data dependent leakage path. For the worst-case data pattern, cell leakage increases by 30 % compared to the 6T cell. The cell leakage can be reduced drastically by using HVT devices in the cell. However, using HVT transistors in the read path of the cell reduces the I_{READ}. The degradation in I_{READ} is further aggravated by the ever increasing V_T mismatch, as well as to process and temperature variations. The time required for the development of the bit-line voltage difference increases with decreasing I_{READ}, which directly increases the read access time.

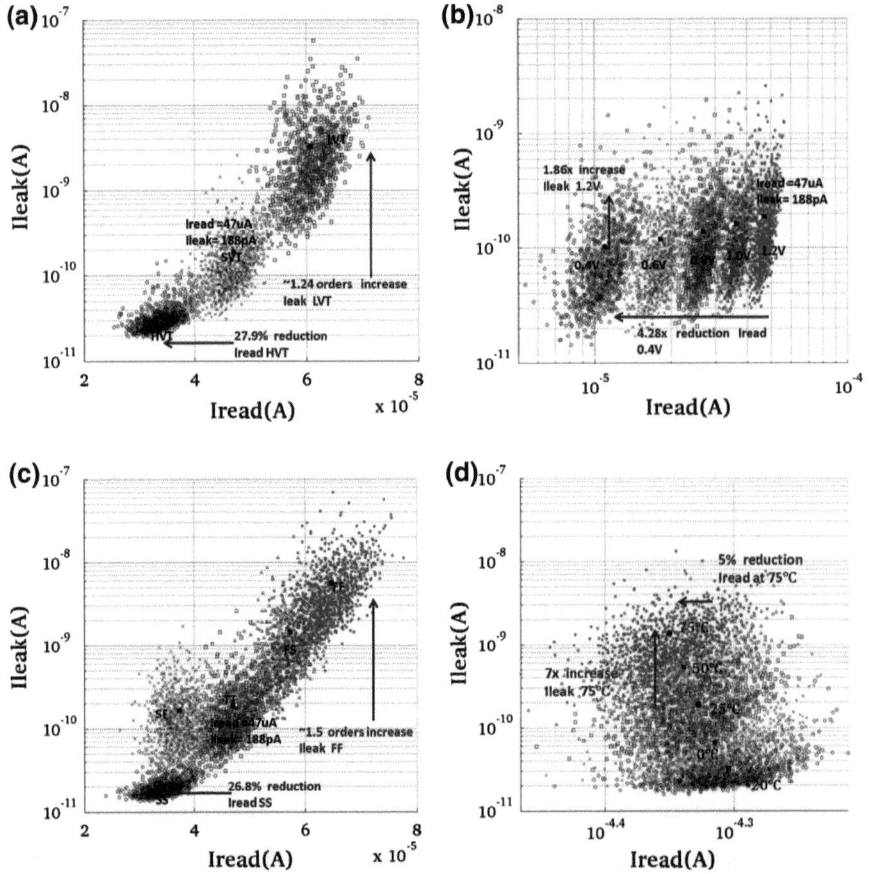

Fig. 2.12 RD8T cell, 65 nm LP technology, minimum sized SRAM cell. **a** Vt variation. **b** VDD variation for standard Vt RD8T cell. **c** Process variation for standard Vt RD8T cell. **d** Temp variation for standard Vt RD8T cell

Fig. 2.13 10T cell (Calhoun et al. 2006)

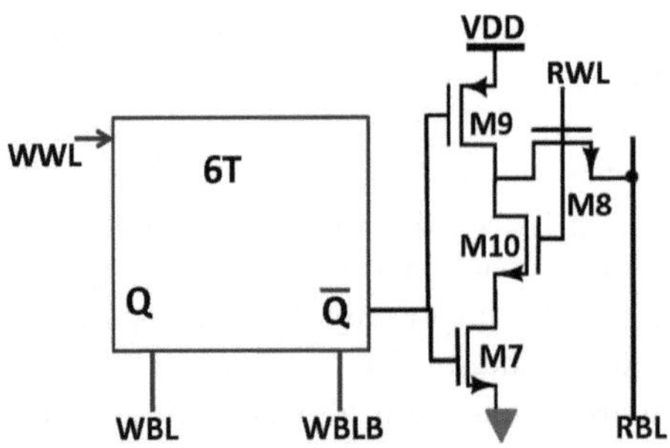

Dual Vt 8T SRAM cell. HVT transistors are used for the 6T part of the cell (the cross-coupled invertors and the write access transistors). This results in a large reduction in the leakage current, as the cross-coupled inverters and the write access

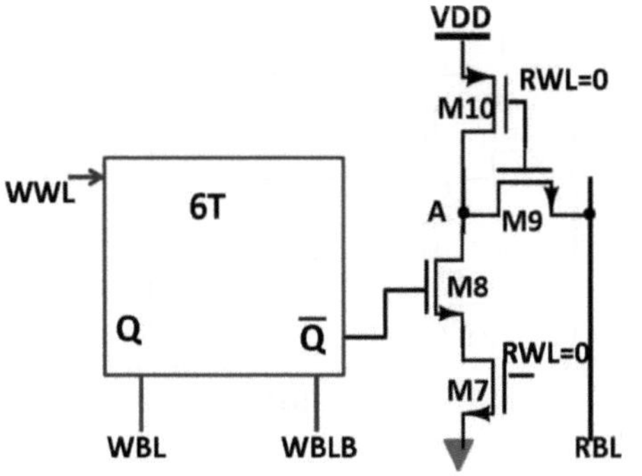

Fig. 2.14 DIL 10T cell (Kim et al. 2007)

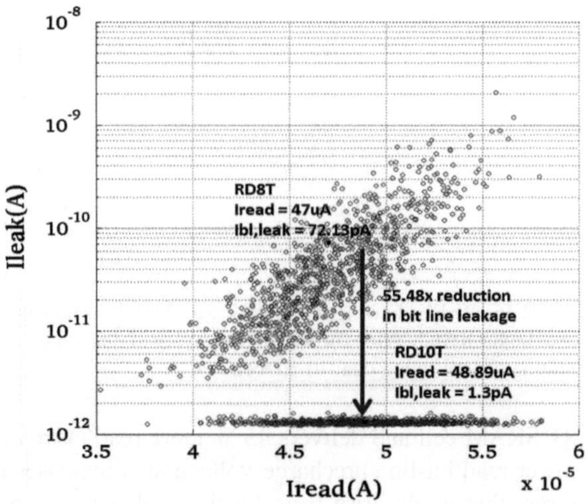

Fig. 2.15 Bit-line leakage reduction with DIL 10T cell (Kim et al. 2007) in 65 nm at VDD = 1.2 V

transistors contribute 70 % of the leakage. The two stacked NMOS transistors (read buffer) determine the read access time. To meet the target performance requirements, SVT transistors are used for the read port.

Reduced Swing Dual Vt 8T SRAM cell (Fig. 2.16). Leakage is further reduced by reducing the read bit-line precharge voltage. In a traditional 6T cell, the bit-line precharge voltage cannot be reduced below VDD-Vt as this increases the risk of read instability. Due to the isolation of the internal storage node from the read bitline, 8T cells do not suffer from this issue, hence a lower read bit-line voltage of 0.2 V is used in this design. The lower drain-to-source voltage reduces the bit-line leakage current because of the reduced drain-induced barrier lowering (DIBL).

The reduced read bit-line precharge voltage (0.2 V) further reduces the leakage current on the read bitline with 3.5× in case of the worst data pattern (Q = "H" for all nonselected cells). Using a low precharge voltage on the read bitline also reduces the dynamic energy consumption discussed in Chap. 5. The dual Vt 8T SRAM cell with 0.2 V read bit-line precharge voltage consumes only 20 % more leakage current compared with the VDD precharged read bit-line single Vt HVT

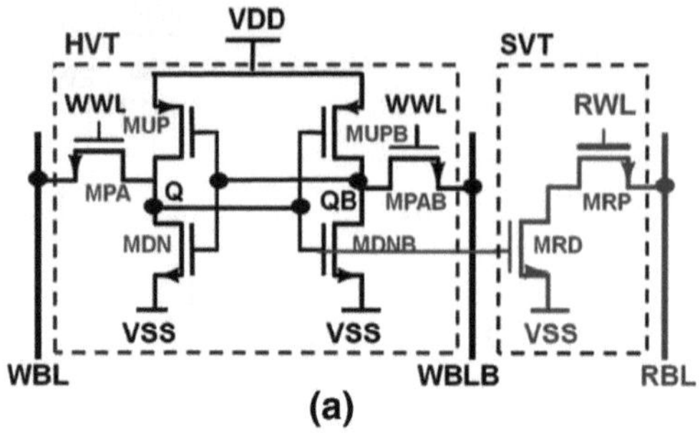

TABLE I
Normalized dual Vt 8T sizing

Pull-up	(W_{MUP} / L_{MUP})	1.25/1
Pull-down	(W_{MDN} / L_{MDN})	1.25/1
Pass-gate	(W_{MPA} / L_{MPA})	1.25/1
Read Pull-down	(W_{MRD} / L_{MRD})	2/1
Read Pass-gate	(W_{MRP} / L_{MRP})	1.25/1

(b)

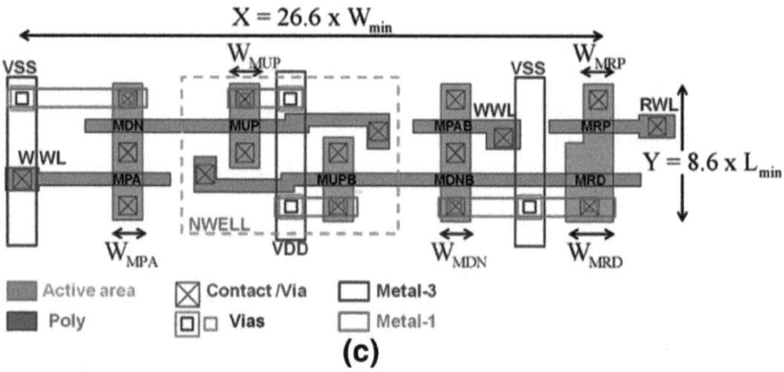

Fig. 2.16 Reduced swing dual Vt 8T SRAM cell. **a** Schematic (HVT write structure and SVT read buffer RB). **b** Table for normalized sizing of dual Vt 8T cell transistors. **c** Layout with power routing of low swing dual Vt 8T cell

8T SRAM cell and delivers 45 % more read current. For a given bit-line swing, a lower read bit-line precharge voltage also increases the resilience to functionality errors that might arise from bit-line leakage. The read bit-line leakage reduction not only reduces the static power consumption, it also improves the read signal. A read failure can occur when the ratio of the read current of the asserted cell to the total leakage current of all the "off" cells on the read bit-line degrades too much due to high leakage currents. It becomes difficult to differentiate between the bit-line discharge caused by the stored data and the bit-line droop because of the leakage current. Figure 2.17 shows Ion/Ioff as function of the supply voltage for different column heights. The dual Vt-8T cell with read bit-line precharged to 0.2 V improves the current ratio with 2.7× compared to a dual Vt-8T cell with read bit-line precharged to VDD. For column height of 256 cells, even for the worst case (FF) process corner and 70 °C the improvement is 1.25× at 0.8 V VDD (Fig. 2.17b). This is because of the leakage mitigation achieved from the reduced DIBL.

4. Dual-Ended Transmission Gate (DETG) Write Cell: WRITE Margin improved (Agarwal et al. 2010)

In the DETG SRAM cell the NMOS access transistors are replaced by full transmission gates (Fig. 2.18). It improves writeability and reduces the WRITE

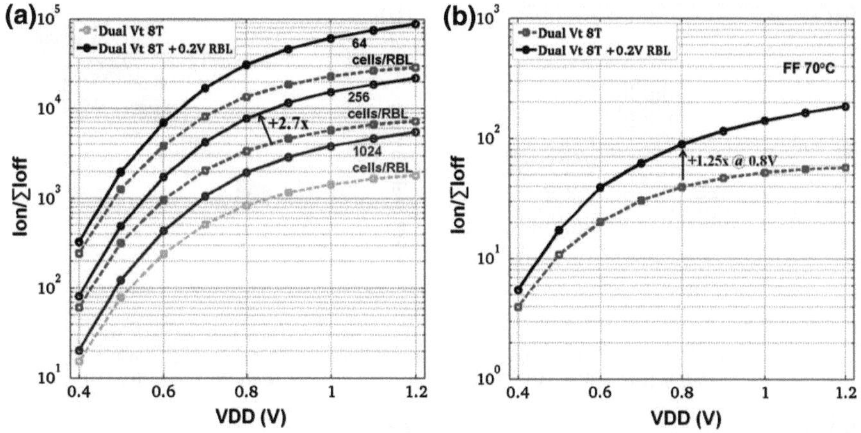

Fig. 2.17 **a** Ion/∑Ioff ratio versus VDD for the different value of column heights. **b** Ion/∑Ioff for the column height of 256 cells (this design) for the worst case FF process corner and 70 °C. Reduced value of RBL reduces the bit-line leakage with the result Ion/∑Ioff is higher with reduced swing dual Vt-8T cell

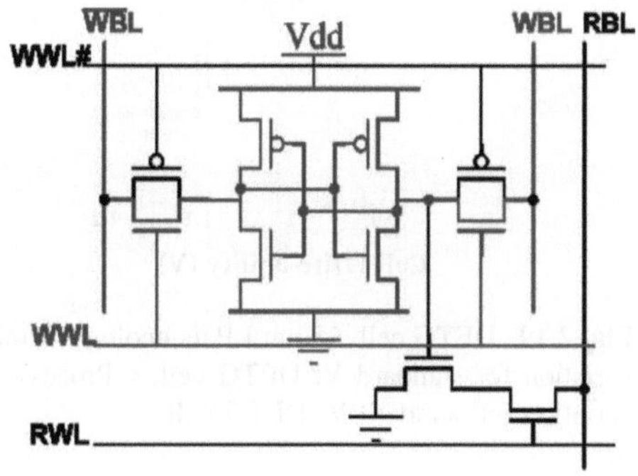

Fig. 2.18 Dual-ended transmission gate (DETG) write memory cell (Agarwal et al. 2010)

access transistor. The WM is 56.3 % more compared to the RD 8T SRAM cell. Figure 2.18 shows WM versus SNMread for different PVT conditions. The READ and WRITE operation is exactly the same as that of RD 8T SRAM cell. The cell symmetry with respect to NMOS and PMOS reduces the effect of the systematic variations and also the redundancy results in averaging out the random variations across the two transistors. Figure 2.19 shows Iread, cell versus leakage for different PVT conditions. The test chip (Agarwal et al. 2010) (register file) based on the DETG cell in 32 nm achieves VDDmin of 0.34 V (Fig. 2.20).

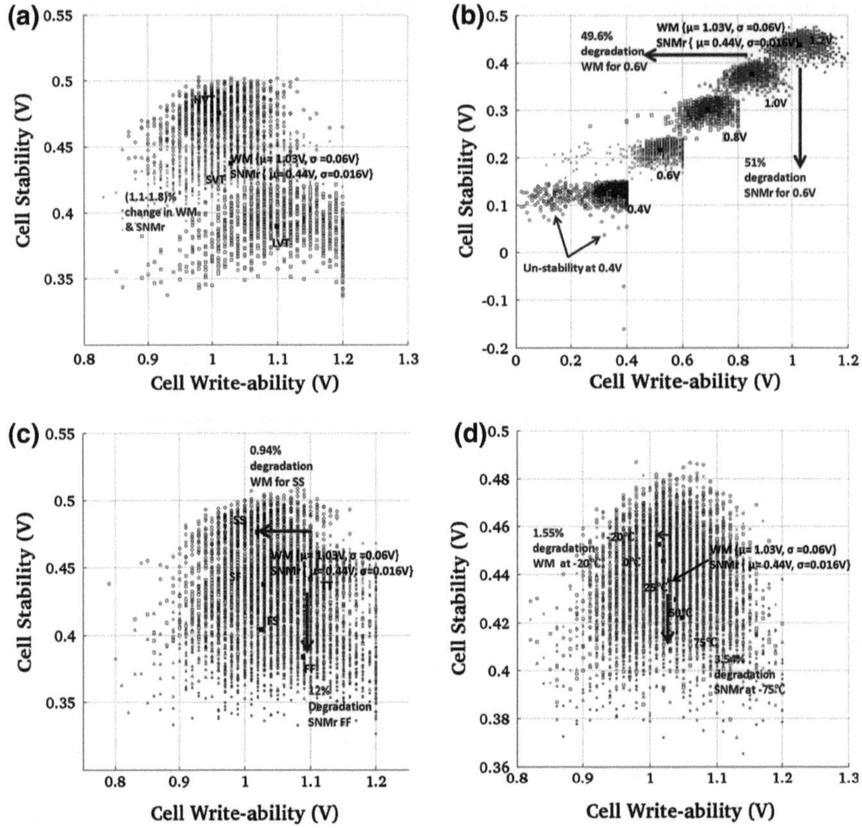

Fig. 2.19 DETG cell, 65 nm LP technology, minimum sized SRAM cell. **a** Vt variation. **b** VDD variation for standard Vt DETG cell. **c** Process variation for standard Vt DETG cell. **d** Temp variation for standard Vt DETG cell

2.2.5 Differential Read Decoupled 8T and 10T SRAM Cells

The read decoupled SRAM cell topologies discussed earlier are single-ended bit cells. There is an inherent loss of common mode noise rejection capability on the bitlines with the single-ended sensing. It is very crucial to ensure a desired level of the noise margin in order to distinguish between genuine bit-line discharge and the read-data droop because of leakage current. In this section read decoupled differential SRAM cells will be discussed.

1. Complementary 10T (CP10T) SRAM Cell (Chang et al. 2009b)

Figure 2.21 shows a read decoupled CP10T SRAM cell. During READ operation WL is activated and VGND is pulled to VSS. W_WL is kept disabled and the internal storage nodes (Q, Qbar) remain isolated from the bit-line. Depending on the storage node information one of the bit-lines starts discharging on the assertion

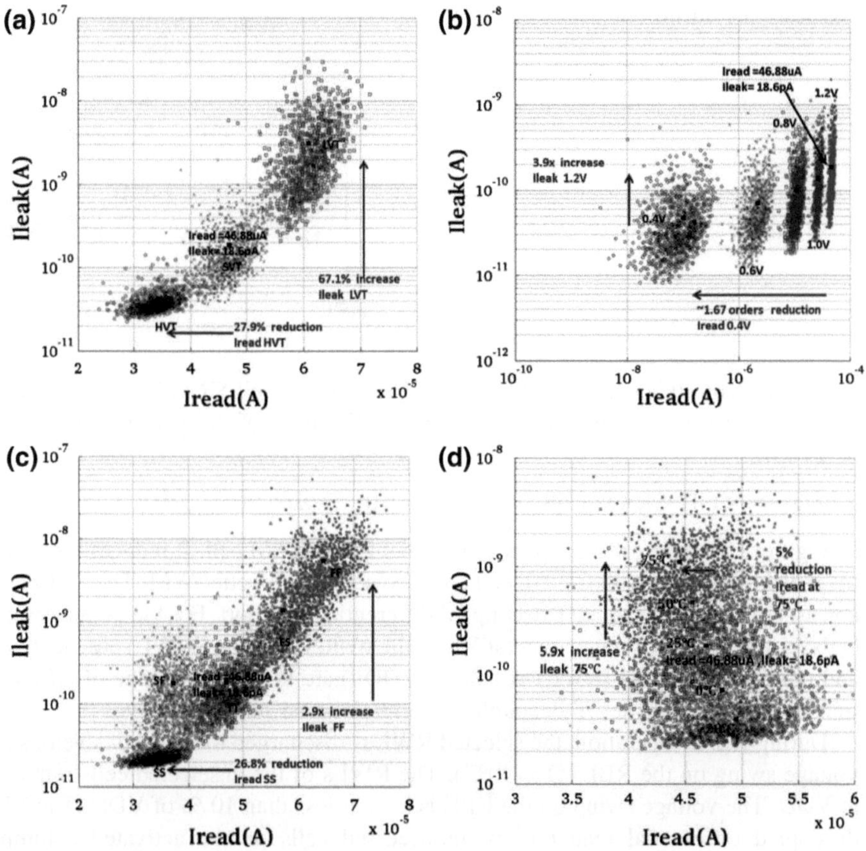

Fig. 2.20 DETG cell, 65 nm LP technology, minimum sized SRAM cell. **a** Vt variation. **b** VDD variation for standard Vt DETG cell. **c** Process variation for standard Vt DETG cell. **d** Temp variation for standard Vt DETG cell

Fig. 2.21 Complementary 10T (CP10T) SRAM cell

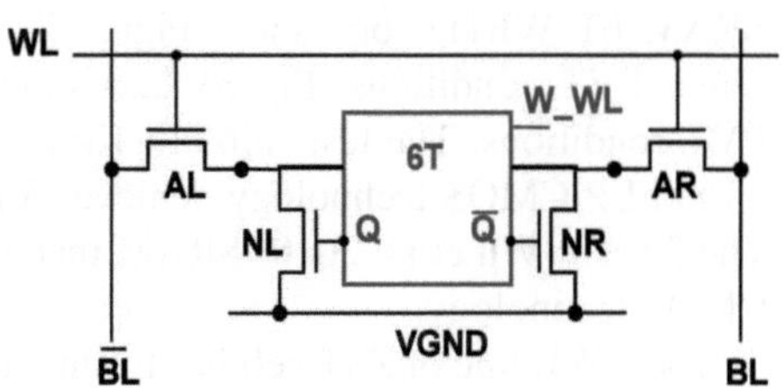

of the WL signal and VGND pulled to VSS. Due to the inverted nature of the sensing the bit-line positions are swapped.

During WRITE operation both WL and W_WL are activated to transfer the write data to cell node from bitlines. Two series access transistors degrade the writeability of the CP10T cell. This results in 44.44 % degradation in the WM.

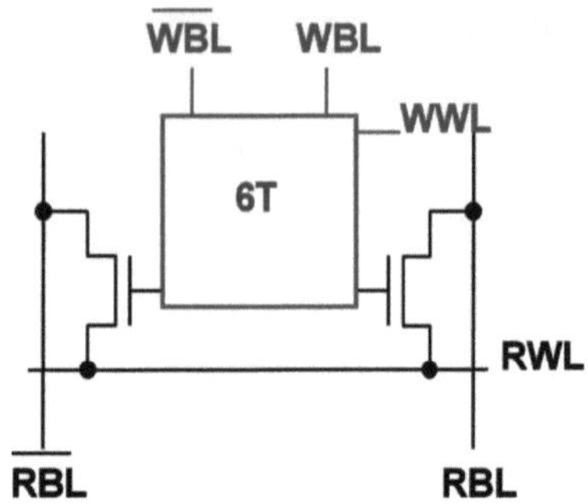

Fig. 2.22 Z8T SRAM cell

The test chip (Chang et al. 2009b) of 32 Kb CP10T cell SRAM macro (Chang et al. 2009b) in 90 nm CMOS achieves VDDmin of 0.18 V operating at 31.25 kHz.

2. Zigzag (Z) 8T SRAM Cell (Wu et al. 2010) (Suzuki et al. 2010)

Figure 2.22 shows a decoupled differential Z8T SRAM cell. It reduces the area overhead associated with CP10T cell and also achieves a better WM. The Z8T cell consists of a 6T cell and a 2T decoupled differential read port. For the un-accessed cells, bitlines are kept precharged at VDD, the write word line (WWL) are inactive and the read word line (RWL) is held at VDD (gate-to-source voltage of NMOS transistors is zero). The 2T decoupled differential read port is inactive.

During READ operation, the selected RWL is discharged to low and develops a voltage swing on the RBL (Q = "H"). The RWLs of the unselected cells remain at VDD. The voltage swing on the RBL is kept at less than 10 % of VDD. The 2T decoupled differential read port of un-accessed cells on the activated column remains in the cut-off region. The potential risk of the bit-line leakage is avoided. The differential read and suppressed BL leakage achieves faster read access. For enhancing the write ability WRITE access transistors are upsized in order to achieve better writeability. The WRITE operation of Z8T cell is similar to the SRAM 6T WRITE operation. Figure 2.23 shows SNMread versus WM for different PVT conditions. Figure 2.24 shows Iread versus leakage under different PVT conditions. The test chip (14 Kb Z8T SRAM macro) (Suzuki et al. 2010) in 65 nm LP CMOS technology achieves VDDmin of 0.5 V operating at 154 MHz. The 32 Kb (Wu et al. 2010) SRAM macro achieves VDDmin of 0.44 V in 90 nm CMOS technology.

The RWL line of Z8T cell has to sink all the discharge current of differential 2T and can result in serious voltage drop thereby impacting the gate-to-source voltage of differential transistors and the Iread, cell. The IR drop on RWL line for the wider word lengths can result in a severe degradation of Iread, cell. With the result Z8T SRAM cell and CP10T structure inhibits its usage for large test arrays. Alternatively, pseudo 8T gated read buffer local architecture (Sharma et al. 2011a) discussed in Chap. 4 enables a differential 8T sensing which can be applied to much larger SRAM arrays and also have a less area overhead.

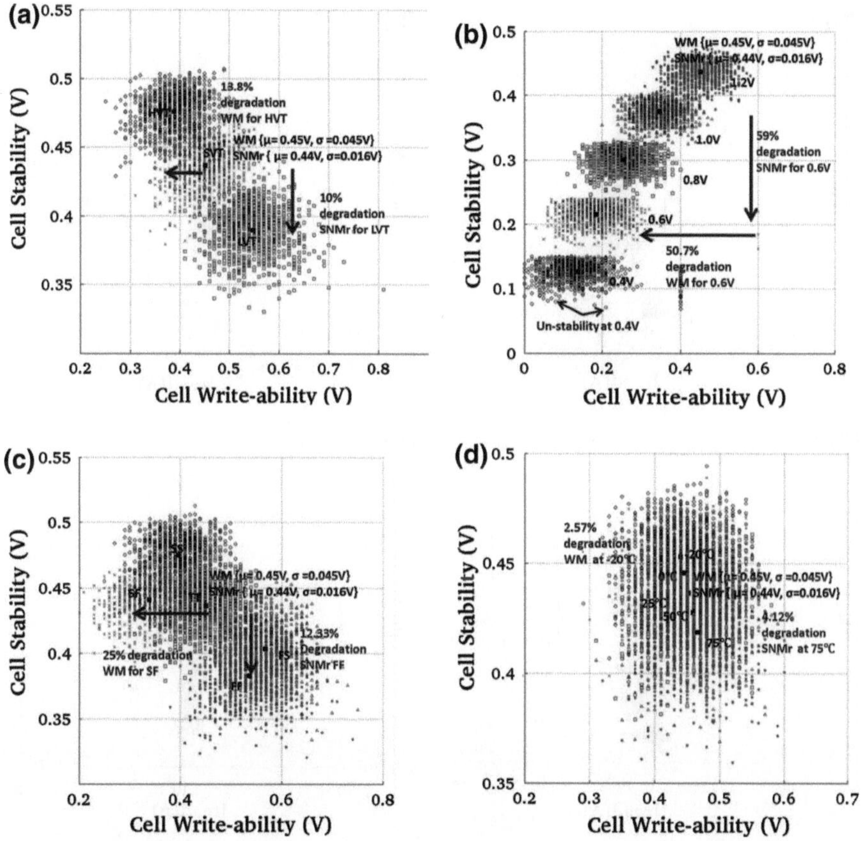

Fig. 2.23 Z8T cell, 65 nm LP technology. **a** Vt variation. **b** VDD variation for standard Vt Z8T cell. **c** Process Variation for standard Vt Z8T cell. **d** Temp variation for standard Vt Z8T cell

2.3 Summary

Table 2.1 shows the comparison of different SRAM cell topologies with reference to the conventional 6T SRAM cell. As it can be observed there is not a single cell topology which can address all the issues like SNMread, WM, half select condition free and also occupies minimum cell area. But the reduced swing dual Vt 8T cell solves most of the issues and is a logical choice for designs in advanced technologies as it avoids read disturbs and allows optimizing the 6T core for write ability. As the 6T core has no impact on memory speed, it can be implemented with slow, low leakage transistors, significantly reducing the standby power consumption. The read buffer current has a large impact on the memory speed, hence the use of fast, low-Vt transistors. This not only improves the nominal read current, but also the variations on the read current thanks to the increased gate-source overdrive voltage. This improvement is most welcome in scaled

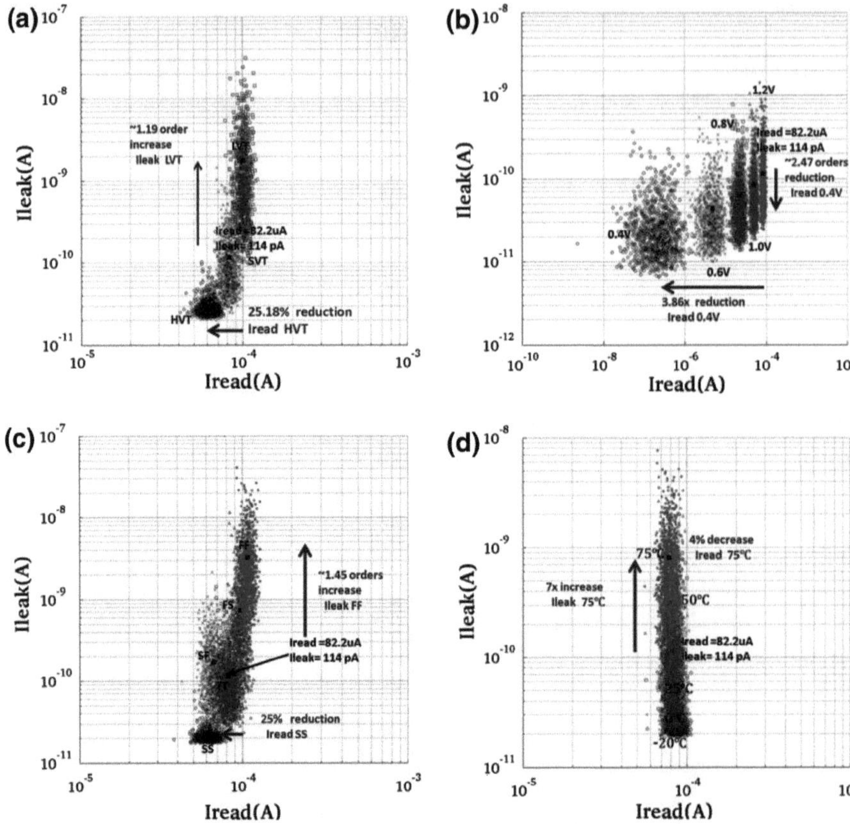

Fig. 2.24 Z8T cell, 65 nm LP technology. **a** Vt variation. **b** VDD variation for standard Vt Z8T cell. **c** Process Variation for standard Vt Z8T cell. **d** Temp variation for standard Vt Z8T cell

designs with lower VDD and higher transistor variations. Additionally, the low precharge voltage reduces the average bit-line discharge energy and improves the Ion/Ioff ratio on the read bit-line.

The D2APT 8T cell does not rely on voltage modulation for higher WMs but it does not offer improved cell stability. The DETG cell offers very high write and cell stability margins but it consists of 10 transistors and also the read sensing is single ended. The Z8T SRAM cell offers better read stability and differential sensing, but there is an inherent problem in its topology as discussed which limits it applicability. The best solution for the variability resilience and low energy lies in combining the cell topology based solution as discussed with the voltage modulation and local architecture modifications of SRAM array are discussed in next chapters.

Table 2.1 Summary of different SRAM cell topologies with reference to SRAM 6T cell

	RSNF7T	D2AP8T	CR8T	RD8T	RSDVt8T	DIL10T	DETG	CP10T	Z8T
Min cell area	1.13×	1.5×	1.3×	1.3×	1.3×	1.8×	…	2.02×	1.34×
#Control signals	3	2	2	2	2	2	3	3	2
Sensing	Single	Differential	Differential	Single	Single	Single	Single	Differential	Differential
VDD min	0.44 V (90 nm)	0.54 V (45 nm)	0.6 V (45 nm)	0.41 V (65 nm)	0.7 V (65 nm)	0.4 V (45 nm)	0.34 V (32 nm)	0.2 V (90 nm)	0.5 V (65 nm)
Read disturb free	Yes	No	No	Yes	Yes	Yes	Yes	Yes	Yes
Write ability	…	High	Low	Same	Same	Same	High	Low	Same
Half-select condition immunity	High	High	Very high	Same	Same	Same	Same	Same	
Leakage	High	Low	High	Low	Low	Low	High	Same	Low

References

A. Agarwal et al., A 32 nm 8.3 GHz 64-entry × 32b Variation Tolerant Near-Threshold Voltage Register File. *Symposium on VLSI Circuits Digest of Technical Papers*, pp. 105–157 (2010)

B.H. Calhoun et al., A 256 k Sub threshold SRAM Using 65 nm CMOS. *Proceedings of IEEE International Solid-State Circuits Conference (ISSCC)*, pp. 628–629, Feb 2006

L. Chang et al., Stable SRAM Cell Design for the 32 nm Node and Beyond. *Symposium on VLSI Circuits Digest of Technical Papers*, pp. 128–129 (2005)

L. Chang et al., An 8T-SRAM for variability tolerance and low-voltage operation in high-performances caches. IEEE J. Solid-State Circuits, **43**, 4, April (2008)

M.F. Chang et al., A Differential Data Aware Power-supplied (D^2AP) 8T SRAM Cell with Expanded Write/Read Stabilities for Lower VDDmin Applications. *Symposium on VLSI Circuits Digest of Technical Papers*, pp. 156–157 (2009a)

I.J. Chang et al., A 32 kb 10T sub-threshold SRAM array with bit-interleaving and differential read scheme in 90 nm CMOS. IEEE J. Solid-State Circuits **44**(2), 650–658 (2009b)

T.H. Kim et al., A High-Density Sub threshold SRAM with Data–Independent Bitline Leakage and Virtual Ground Replica Scheme. *Proceedings of IEEE International Solid-State Circuits Conference (ISSCC)*, pp. 330–331, Feb 2007

V. Sharma et al., A 4.4 pJ/Access 80 MHz, 128 kbit variability resilient SRAM with multi-sized sense amplifier redundancy. IEEE J. Solid-State Circuits, **46**, 10 (2011a)

V. Sharma et al., 8T SRAM with Mimicked Negative Bit-lines and Charge limited Sequential Sense Amplifier for Wireless Sensor Nodes. *Proceedings of IEEE European Solid-State Circuits Conference (ESSCIRC)*, pp. 531–534, Sept 2011b

T. Suzuki et al., 0.5 V, 150 MHz, Bulk-CMOS SRAM with Suspended Bit-Line Read Scheme, *Proceedings of IEEE European Solid-State Circuits Conference (ESSCIRC)*, pp. 354–357, Sept 2010

K. Takeda et al., A read-static-noise-margin-free SRAM cell for low-VDD and high-speed applications. IEEE J. Solid-State Circuits **41**(1), 113–121 (2006)

K. Utsumi et al., A 65 nm low power CMOS platform with 0.495 μm^2 SRAM for digital processing and mobile applications. *Proceedings of IEEE Symposium VLSI Technology*, pp. 216–217 June 2005

J. Wu et al., A Large σV_{TH}/VDD Tolerant Zigzag 8T SRAM with Area-Efficient Decoupled Differential Sensing and Fast Write-Back Scheme. *Symposium on VLSI Circuits Digest of Technical Papers*, pp. 101–102 (2010)

M. Yabuuchi et al., A 45 nm low-standby-power embedded SRAM with improved immunity against process and temperature variations. *Proceedings of IEEE International Solid-State Circuits Conference*, pp. 326–327, Feb 2007

M. Yabuuchi et al., A 45 nm 0.6 V Cross-Point 8T SRAM with Negative Biased Read/Write Assist, *Symposium on VLSI Circuits Digest of Technical Papers*, pp. 158–159 (2009)

Chapter 3
Adaptive Voltage Optimization Techniques: Low Voltage SRAM Operation

The minimum supply voltage for SRAM cell is limited by write failures (write-ability) or read disturb failures (cell stability). In the previous chapter, various SRAM cells are discussed which offer better variability resilience compared to the SRAM 6T cell and enable low VDD operation. SRAM 6T cell functionality is highly dependent on the supply voltage. The voltage optimization can impact the write failures and the read disturb failures significantly. In this chapter, dynamic voltage optimization techniques, are studied for realizing low VDD operation with the conventional SRAM 6T cell while maintaining sufficient READ/WRITE margins. The goals of this chapter are as follows

1. Implementation details of different dynamic voltage optimization techniques.
2. A comparitative analysis of various voltage optimization based assist techniques to improve the variability resilience of SRAM 6T cell.
3. An overview and comparison of hybrid voltage optimization techniques like configurable write assist, Crosshair SRAM, mimicked negative bit-line (MNBL) technique, and compounded differential VSS (CDVSS) bias technique.

3.1 Introduction

Figure 3.1, illustrates a SRAM 6T cell, with a latch transistor (back to back inverters holding the storage node values) and the pass transistors providing the access to the storage nodes from the bit-lines. During WRITE operation (writing "H" at left node) the voltage at right node needs to be pulled below the trip point of the inverter that drives the left node when the word line is asserted. The increased strength of the right pull up PMOS transistor or the decreased strength of

V. Sharma et al., *SRAM Design for Wireless Sensor Networks*,
Analog Circuits and Signal Processing, DOI: 10.1007/978-1-4614-4039-0_3,

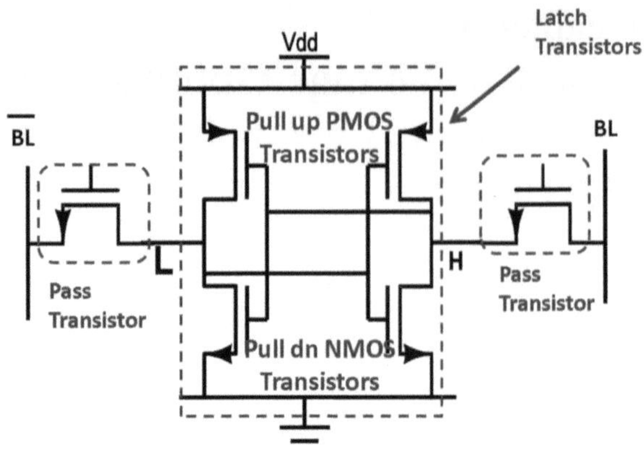

Fig. 3.1 Conventional SRAM 6T Cell

the right pass transistor due to the process variations impedes the discharge process. Furthermore, process variations also reduce the trip point of the left inverter, resulting in write failure. Voltage optimization reduces the strength of the pull up PMOS transistor or increases the strength of pass transistor and avoids the write failure.

During READ operation when the word line is asserted it exposes internal left node (storing "L") to the VDD pre-charged bit-line. Due to the process variations the strength of the left pull down NMOS transistor decreases and the left pass transistor increases. The charge sharing between the bit-line and the internal node increases the probablility of the data flip and might result in a read upset failure. This can be avoided by increasing the strength of the latch transistors or by reducing the strength of the pass transistors by applying an appropriate voltage bias.

Various voltage optimization techniques for improving the variability resilience of SRAM 6T cell can be broadly classified into 2 categories

1. Techniques that impact the storage node information by modulating the strength of the latch transistors of SRAM cell.
2. Techniques that limit the impact of the bit-line voltage by modulating the strength of the pass transistors.

Voltage optimization based WRITE assist technique

Modulating latch strength (reducing)	VDD lowering, VSS raising
Modulating pass transistor strength (increasing)	WL boosting, Negative Bit-line

Voltage optimization based READ assist technique

Modulating latch strength (increasing)	VDD raising, VSS lowering
Modulating pass transistor strength (reducing)	WL suppression, Bit-line voltage reduction

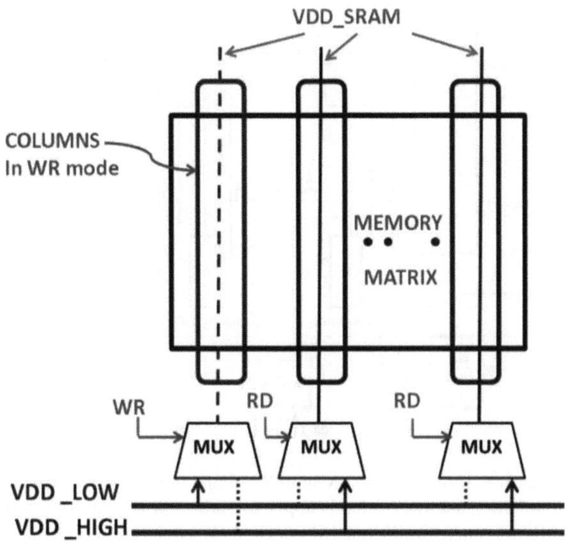

Fig. 3.2 Dual column supply voltage optimization (Zhang et al. 2006). VDD mux provides a choice between two power supplies (VDD_HIGH and VDD_LOW). Implementation: 70 Mb SRAM macro in 65 nm

3.2 WRITE Assist Techniques

A. **Weak PMOS pull up transistor based techniques: SRAM VDDcell (power line) voltage optimization**

1. **Dual Power Supply**

Zhang et al. (2006) proposes a dual power supply scheme in order to achieve higher write margins. The SRAM cell is made unstable during the WRITE operation by lowering its supply voltage. Figure 3.2 shows a column based dual supply implementation scheme. During WRITE operation, the column that are being written remains on the low supply voltage (VDD_LOW), allowing easy overwriting of the cells. The value of the low supply voltage (VDD_LOW) is taken such that there is no data retention issue for the unactivated cells on the selected column. Whereas the columns which are not being written are powered with higher supply (VDD_HIGH). The 2 power supplies (VCC_hi and VCC_lo) are provided externally.

Dual column supply implementation involves switching of high column capacitances, column height of 128 cells (Zhang et al. 2006) during the WRITE operation. This results in increasing the dynamic energy consumption and the WRITE start operation latency, especially for the higher column heights (256–1024 cells). This problem is addressed in (Sharma et al. 2010). Figure 3.3 shows the implementention of fine grained VDDcell switching (local blocks).

The height of the local block is 8 cells which is 1/64 of the height of an entire column of the memory matrix. This implementation further optimizes the energy consumption associated with the switching of the supply capacitances by

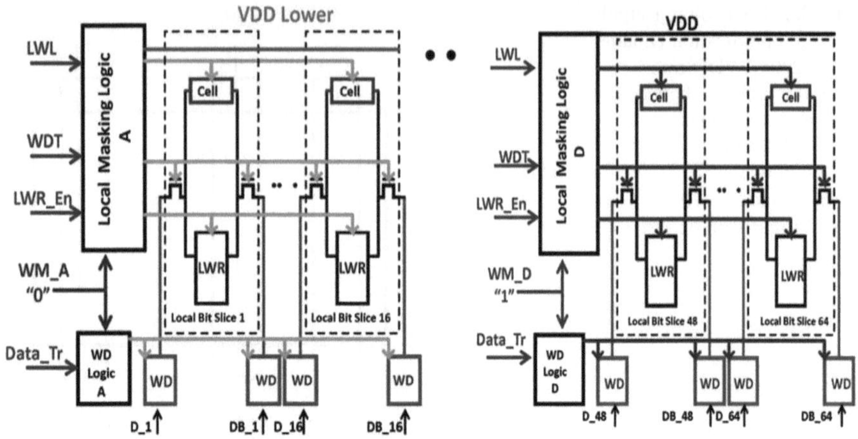

Fig. 3.3 Dual local block supply voltage optimization (Sharma et al. 2010). Local masking logic implements a fine grained supply lowering for a quarter of 64 bits word. Implementation: 128 Kb SRAM macro in 90 nm LP

providing a selective VDDcell switching for the certain bits of the word length. The data correlation is exploited and the VDDcell switching for the certain set of bits of a data word is prevented. During WRITE operation only the power supply of a quarter (8 cells $\times$ 16 cells) of an activated local block (8 cells $\times$ 64 cells) is switched to VDD Lower. The cell supply of the quarters where there is no requirement for the WRITE operation remains at VDD. The selective VDD lowering information is provided by the external pins.

2. **Charge sharing for VDDcell reduction—Capacitive Write Assist**

Capacitive write assist technique enables VDD lowering during the WRITE operation without requiring an extra external power supply. Ohbayashi et al. (2007) realizes an easy WRITE operation by slightly lowering the VDD of the SRAM cells to be written with the help of a capacitive assist circuit. Figure 3.4 shows the capacitive write assist circuit. During an idle state all the P-Trs and the N-dr are enabled and the voltage level of the SRAM array (VDD_SRAM) is VDD and the VDD_DN (metal layer routing) is pulled down to VSS. The SRAM array supply voltage lines are isolated from the VDD_DN, as all the N-Trs are disabled.

During WRITE operation WC_EN[i] of the activated matrix column (MC[i]) disables P-Tr[i] and WR_EN disables N-dr. With the result VDD_SRAM and VDD_DN are floating and N-Tr[n] shorts VDD_SRAM and VDD_DN. This results in charge sharing between VDD_SRAM and the VDD_DN lines which lower the supply voltage, proportional to the capacitance ratio of the VDD_SRAM and the VDD_DN line. This lowering of the VDD_DN line voltage weakens the pull up transistors of the memory cell and the write-ability of the SRAM cell is improved.

Yabuuchi et al. (2007) proposes a divided VDD array (VDD_SRAM) scheme (fine grained implementation of the VDD lowering with the capacitive write assist

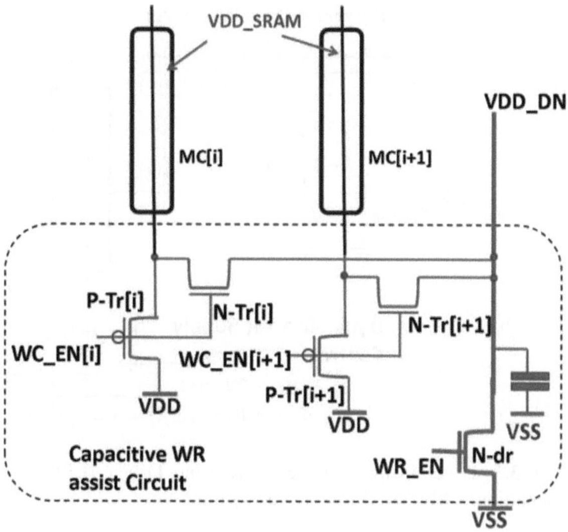

Fig. 3.4 Capacitive write assist technique (Ohbayashi et al. 2007). The charge sharing lowers the array supply voltage. Implementation: 8 Mb SRAM macro in 65 nm

technique). The VDD lowering with the capacitive write assist technique depends on the ratio of an array capacitance (VDD_SRAM) and a dummy wiring capacitance (VDD_DN). The VDD_DN line is divided into 8 segments. This results in 30 % lowering of the SRAM array voltage during the WRITE operation.

The extra metal resources required for routing of the VDD_DN lines lowers the SRAM cell density. Second, the floating power lines during WRITE results in the data retention problem for the unactivated cells sharing the floating power lines.

3. **VSS (ground raising)** (Yamaoka et al. 2004)

Alternative to the VDD lowering, VSS raising can also be used to improve the write-ability of the SRAM cells. Yamaoka et al. (2004), implements VSS row based voltage optimization for improving the write-ability of the SRAM cells. During WRITE operation the source line of the SRAM cells (VSS_SRAM) is disconnected from the VSS. The source line voltage is allowed to float. The raised (VSS_SRAM) decreases the drive strength of the PMOS transistors of SRAM cells and results in improved write-ability of the cell.

B. **Strong NMOS pass transistor based techniques: word line (Vwl) voltage optimization**

4. **Level programmable word line driver** (Hirabayashi et al. 2009)

Hirabayashi et al. (2009) proposes a dual power rail SRAM. The bit-line pre-charge circuitry and the other peripheral circuits operate at VDD_LOW (0.8 V). Whereas, the level programmable word line drivers (LPWD) and the SRAM array are supplied with both VDD_HIGH (1.0 V) and VDD_LOW (0.8 V). An intermediate word line voltage level (Vwl) between VDD_HIGH and VDD_LOW is generated by a LPWD. Figure 3.5 shows dual power supply with

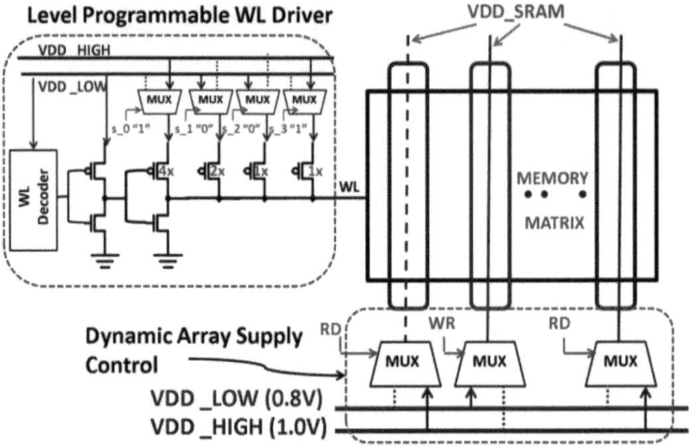

Fig. 3.5 Level Programmable Word Line Driver (LPWD) (Hirabayashi et al. 2009). Implementation 2 Mb SRAM macro in 40 nm

LPWD. In a LPWD, the WL pull up PMOS transistors organized in a binary manner (px4-px1) so that the source node of each pull up PMOS transistors can be independently connected to VDD_HIGH or VDD_LOW. The gate inputs of pull up PMOS transistors are programmable with the signals s_0–s_3. A combination of s_0–s_3 results in different source voltages for the pull up PMOS transistors for an asserted WL. As a result there are some VDD_LOW powered PMOS devices and some VDD_HIGH powered PMOS devices. The current from VDD_HIGH to VDD_LOW flows through the PMOS transistors and the selected WL is biased to a level determined by the conductance ratio of PMOS transistors powered by VDD_HIGH and VDD_LOW. This is how the selected WL is programmed from VDD_LOW to VDD_HIGH. But the drawback is that by using PMOS dividing, DC current is consumed. The amount of DC current consumed depends on the Vwl to be generated. The maximum current flows only when the Vwl selected is midway between VSM and VDD.

5. **WL Boosting with Charge Pump Circuit**

The use of a second higher supply for the word line boosting adds to a platform cost and require extra metal resources. This problem is solved by using single charge pump circuit (Sinangil et al. 2011). The WL signals at the beginning of the WRITE operation are at VDD. The charge pump circuit is turned ON by a short pulse boost signal, generated from a negative edge of the clock. The value of the boosting capacitance is selected such that the 100 mV boosting voltage is achieved. However, there is an issue of an increased area overhead and increased energy consumption associated with the on-chip boosting circuit. The charge pump power overhead is proportional to the required boosting ratio for a given load current. This problem is remedied by Raychowdhury et al. (2010) at the cost of increased test complexity. It proposes a write-ability sensor which autonomously

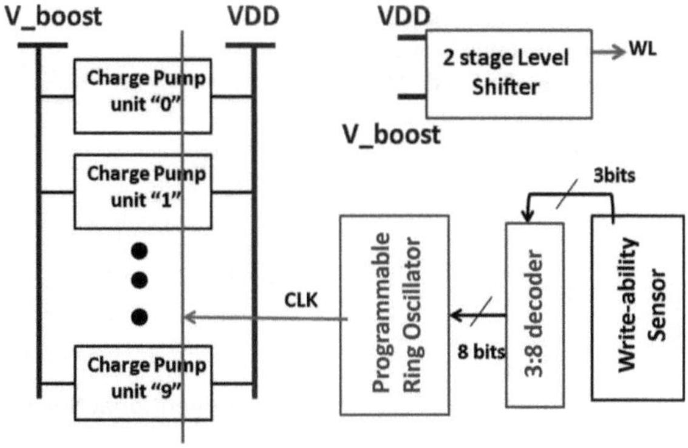

Fig. 3.6 Write-ability sensor tunes the clock frequency for adjusting the boosting ratio for different PVT conditions and further load current is reduced by performing 2 step transitechnique.During the READ operation for the asserted WL from VSS → Vboost (supplied by charge pump unit)

adjusts the boosting ratio as a function of process (P), voltage (V), and Temperature (T) (Fig. 3.6). The 2 stage level shifter minimizes the dynamic load current by performing the boosting transition in 2 stages. The digital readout from the majority detector (write-ability sensor) is indicative of the writeability of a typical cell. And is used to program the frequency of the clock (increasing for high boosting ratio and decreasing for the low boosting ratio) of a charge pump circuit. The boosting ratio is programmed from 30 % to 60 % by adjusting the frequency of the clock.

6. **Negative Bit-line Technique**

The reduced read SNM for the half selected SRAM cells on the same word line puts an upper limit on the value of word line boosting applied. The word line boosting scheme relative to the VDDcell reduces the cell stability (read SNM) of the half selected SRAM cells on the same word line (for the classic decoder architecture). Alternatively bit-line biasing can also be used to increase the strength of NMOS transistor. The write access transistors strength can also be increased by lowering the VSS side of the bit-line to a negative voltage. The VSS side of the bit-line pulled to a negative voltage increases the gate-to-source voltage and the drain-to-source voltage of the pass transistor (Mpr), (Fig. 3.7).

This accelerates the discharge process and the write failures are avoided. The negative bit-line bias does not directly impact the half select bits. The word lines are not activated for the unselected cells in an activated column. Therefore, the increase in the gate-to-source voltage of the access NMOS transistors for the unselected cells is lower than their threshold voltage. In other words, the read SNM with negative bit-line is not degraded.

Nii et al. (2008) implements the negative bit-line technique for expanding the write margin. Self-determining circuitry for the bit-lines adjusts the timing of forcing one of the bit-lines to a negative bias. Capacitance in the circuitry is adjusted to determine the magnitude of the negative bias on the bit-lines. An excessive negative bit-line bias can result in data flips for the inactivated cells on a selected column. Therefore, the magnitude of the applied negative bias on the bit-lines is limited to −0.2 V.

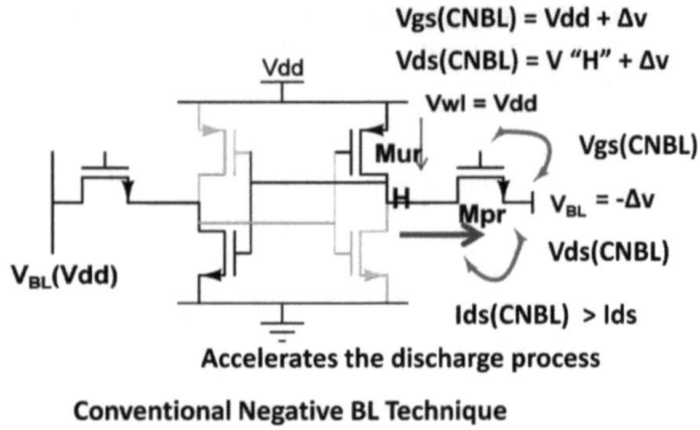

Fig. 3.7 CNBL technique: increases the strength of Mpr by pulling the VSS side to a negative voltage

The problem of an unexpected data flip for the inactivated cells on a selected column is remedied with the use of a cross point 8T-SRAM (Yabuuchi et al. 2009). In a cross point 8T-SRAM, the series access transistors alleviate the internal data flipping. The fabricated 1 Mb SRAM macro using cross point 8T-SRAM with negative bit-line technique in 45 nm achieves correct write functionality at 0.6 V. There is a 100 mV reduction in the operating voltage compared to the (Nii et al. 2008).

7. **Adaptive Negative Bit-line Technique**

The conventional negative bit-line schemes are not suitable for the compilable SRAM. First, the circuit optimized for a fixed array configuration is not good for some other configurations. In conventional negative-bootstrap circuit, the bootstrap capacitance and the timing of activating the signal must be optimized for a certain fixed count of SRAM cells on a bit-line. This makes it unsuitable for the compilable SRAMs. The negative boost applied can be too high and can result in a reliability problem, and on the other hand if too low can result in an insufficient write margin.

This is remedied by using an adaptable constant negative bit-line circuit (Fujimura et al. 2010). Figure 3.8 shows the implementation of a configurable constant negative-level write buffer. The bootstrap circuit automatically adjusts the bit-line bias to an optimized constant-negative-level for the different array configuration of 4–512 cells/BL. The charge stored in a bootstrap capacitor Boost_Cap is automatically controlled depending on the number of rows. The additional charge Δq stored in Boost_Cap is proportional to the increase in the bit-line capacitance. The charge Δq stored on a bootstrap capacitor is equal to the product of the drain current of transistor N_B and the additional time Δt required to pull down the node V_n by N_B.

The time Δt is generated by the Bit-line capacitance monitor. The constant-negative bit-line write buffer adaptively sets Δq to be proportional to ΔCbl with a help of bit-line capacitance monitor. The SRAM macro of 1.25 Mb fabricated in 32 nm achieves minimum operating voltage of 0.7 V with the application of −0.15 V bias voltage.

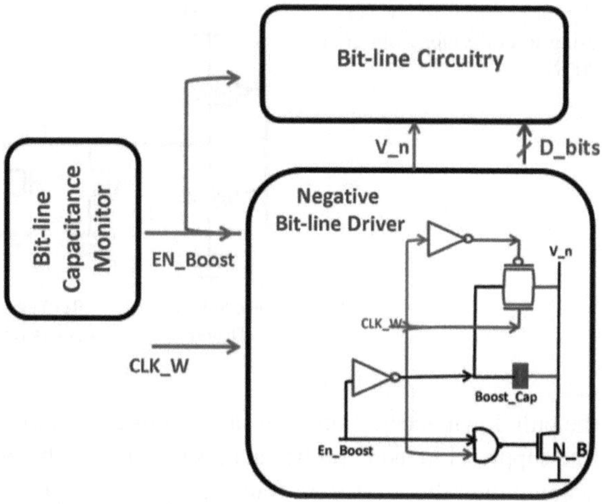

Fig. 3.8 Configurable Constant-negative-level write buffer (Fujimura et al. 2010). The bit-line capacitance monitor connects the replica bit-lines and the capacitance becomes 2Cbl. The amount of charge stored on Boost_Cap is automatically controlled depending on the size of the memory array with the help of bit-line capacitance monitor. Implementation: 1.2 Mb SRAM macro in 32 nm

3.3 READ Assist Techniques

A. **Strong latch transistor based techniques: SRAM Power Line (VDDcell) and the source line (VSScell) Voltage Optimization**

1. **Increasing VDDcell for improving the cell stability**

Zhang et al. (2006), (Yamaoka et al. 2004) and Hirabayashi et al. (2009) as discussed in the WRITE assist technique section, proposed a dual rail SRAM array in which option for two supplies (VDD_high and VDD_Low) is provided. During the READ operation the VDDcell of the activated columns of matrix are switched to higher VDD, thereby increasing the strength of the latch transistors and improving the read SNM.

2. **Cross Point 8T SRAM with Negative VSS**

Yabuuchi et al. (2009), proposed Cross Point 8T SRAM with negative VSS in order to improve cell stability and I_{READ}. During READ operation the column's VSS is negative biased. As a result, the enlarged SRAM cell bias improves the static noise margin (SNM). The application of negative bias for VSS increases the gate-to-source voltage of the access and driver transistor and improves the access speed.

B. **Weak pass transistor based techniques: word line suppression based voltage optimization**

3. **Word line suppression using replica access transistor (RAT) (Ohbayashi et al.** 2007)

The word line voltage is slightly reduced to improve the cell stability. The reduced word line voltage degrades the drive strength of the access transistor. This improves

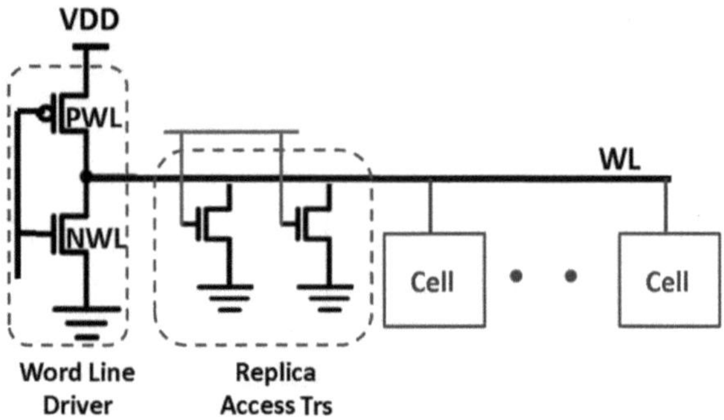

Fig. 3.9 Word line driver using RATs (Ohbayashi et al. 2007)

the pull down/access transistor ratio. Figure 3.9 shows the implementation of word line suppression using replica access transistor. This implementation also takes into account inter die variations. It uses replica access transistors (RATs) that have same physical topology as the access transistor. For the lower values of Vt, the RATs lower the WL voltage more compared to the higher values of Vtn. This is how it achieves a balance between the read SNM and read current.

4. **Fine word line suppression using gate controller with RAT (Yabuuchi et al.** 2007)

The word line voltage lowered by multiple pull down NMOS transistors (Ohbayashi et al. 2007) is not able to track asymmetric process corners for e.g. *fast nmos slow pmos* accurately. The word line voltage level depends on the Vt of a replica access transistor and that of a logic transistor of the word line driver. With the result a unnecessary extra low word line voltage (Vwl) can result in severe degradation of the read current. This problem is remedied by introducing an additional passive resistance element implemented using N-type polysilicon gate (Yabuuchi et al. 2007), Fig. 3.10. The VDD level of the word line driver is determined by the ratio of the resistance R0 and replica access transistor (RAT). The gate length in the *slow nmos slow pmos* process corner becomes longer and the value of resistance decreases, thereby resulting in relatively higher VDD for the word line driver compared to the situation without added resistance. This is how degradation in the cell read current is avoided.

5. **Resistance variation tolerant word line suppression (Nii et al.** 2008)

The word line bias as proposed in (Yabuuchi et al. 2007) is strongly dependent on variation of the resistance (R0). It is difficult to achieve constant word line suppression voltage due to the variation in the resistance value. Variation in the resistance value of R0 maps into variation in the applied word line bias. This problem is remedied by (Nii et al. 2008). The proposed circuitry makes use of an additional resistance R1 as shown in Fig. 3.11 between node Na and the replica access transistor. The variation in the value of an additional resistance inserted counteracts the variation in the value of R0. With the result the sensitive dependence of Na on the variation of the resistance value is suppressed. The variation in

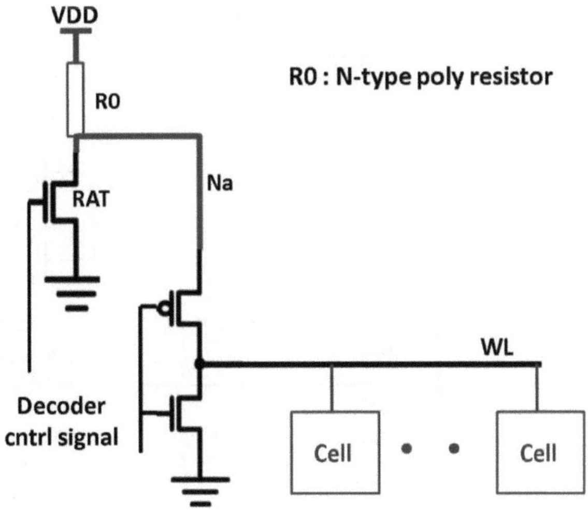

Fig. 3.10 Word line driver using gate controller with RAT (Yabuuchi et al. 2007)

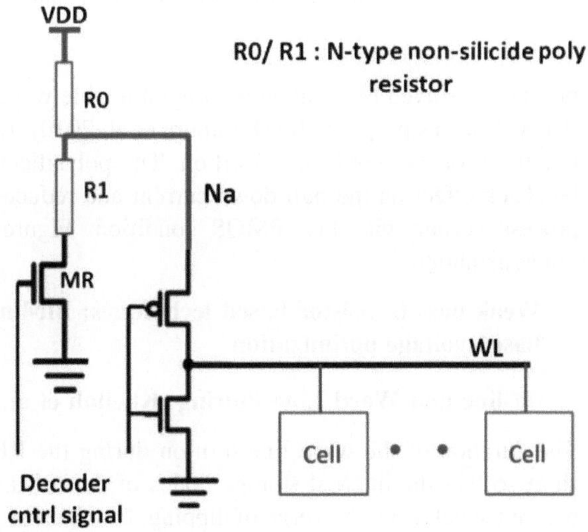

Fig. 3.11 Word line suppression circuitry for read assist. The insertion of additional resistance R1 realizes expected word line voltage (Vwl) of 0.8 × VDD power for a wider range (Nii et al. 2008)

the value of an applied word line bias is suppressed to 5 mV compared to 124 mV with ±30 % variation in the resistance value (Nii et al. 2008). Figure 3.11 shows the word line suppression circuitry, in which the replica transistor (MR) and the n-type poly-silicon resistance (R0, R1) determine the word line bias Na.

6. **Level Programmable Word line driver** (**Fujimura et al.** 2010)

The circuit using word line voltage suppression based read assist technique provides a reduced voltage to the source of pull up PMOS transistor. It decreases the overdrive voltage of a pull up transistor resulting in a slow rise time of WL. This

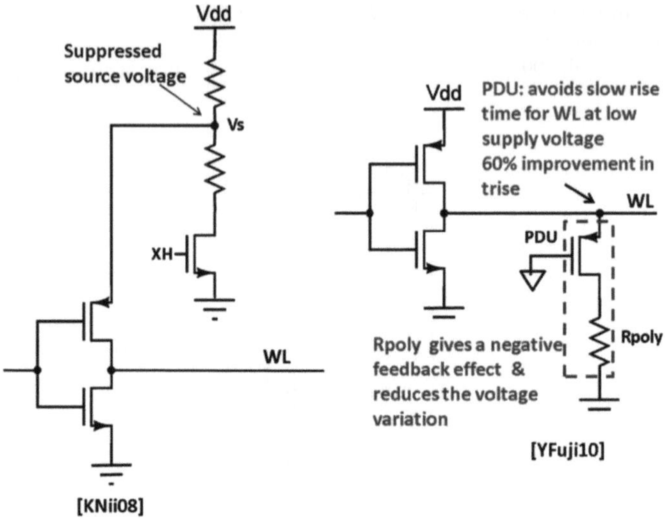

Fig. 3.12 Level word line driver for single supply (Fujimura et al. 2010)

problem is solved by using level programmable word line driver for single supply (LPWD-SS) as proposed by (Fujimura et al. 2010). LPWD-SS provides a superior rise time for the word line assertion. The polysilicon resistance gives a negative feedback effect on the pull down current and reduces the voltage variation at the process corner with fast PMOS condition. Figure 3.12 shows the LPWD-SS implementation.

C. **Weak pass transistor based techniques: Bit-line suppression based voltage optimization**

7. **Bit-line and Word Line Pulsing (Khellah et al.** 2006)

The duration of the word line turn-on during the READ operation is made very short so that the internal storage nodes of the cell are isolated from the bit-lines before the cell is on the verge of flipping. But the word line turn-on duration is long enough to ensure the minimum bit-line differential voltage required for the correct sensing by a sense amplifier. It also utilizes the reduced bit-line pre-charge voltage by using MOS diodes or a source follower. The pre-charged bit-lines are pulled down (100–300 mV) before the word line is asserted. The lower value for the bit-line voltage reduces the amount of charge shared between a bit-line and an internal cell storage node. This is how the probability of data flip is reduced. But, the lower bit-line pre-charge voltage increases the risk of data flips and also degrades the cell read current. Also the determination of an optimum duration for the word line signal is extremely difficult task for the advance technology nodes where the impact of intra die variations is getting more prominent.

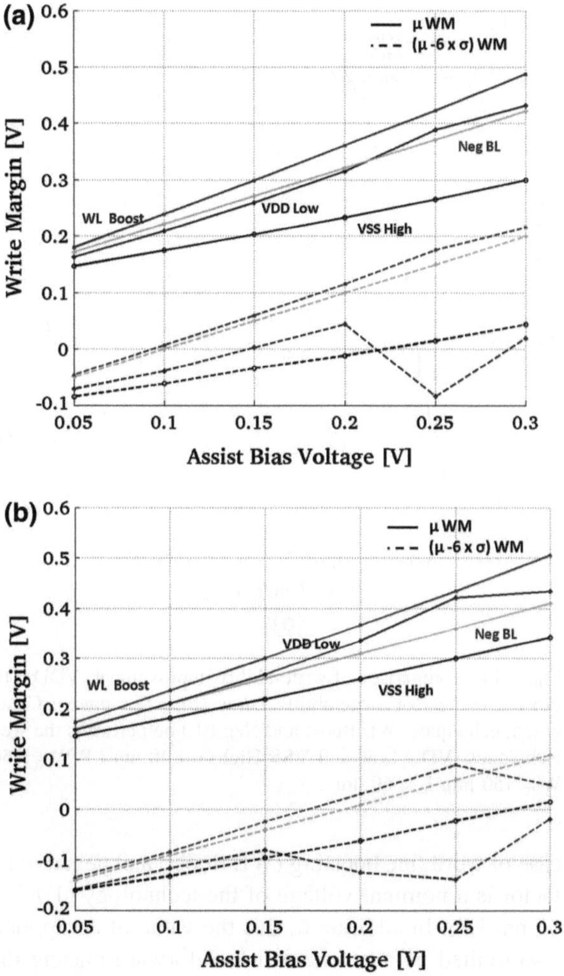

Fig. 3.13 Write Margin versus assist bias voltage Δv for minimum sized HVT transistors based SRAM 6T cell. Higher the values of write margin better the write-ability. VDD = 0.8 V, Temp = 25 °C, and the worst process corner for write-ability (*slow nmos fast pmos*) (**a**) 40 nm LP, (**b**) 65 nm LP

3.4 Comparative Analysis

This section discusses about the functional effectiveness, performance, and energy consumption analysis of various voltage optimization techniques discussed earlier.

1. **Functional Effectiveness**

The change in write margin with respect to the change in an applied bias voltage for a given assist technique is used as a metric for the quantitative comparison of the assist techniques. Figure 3.13 shows the write margin (WM) versus the applied assist bias voltage. In order to account for the impact of the Vt variations on a mean value, 6σ shift in the values of mean WM is also plotted. There are different limiting factors that constrain the value of an applied voltage bias. For e.g. in the

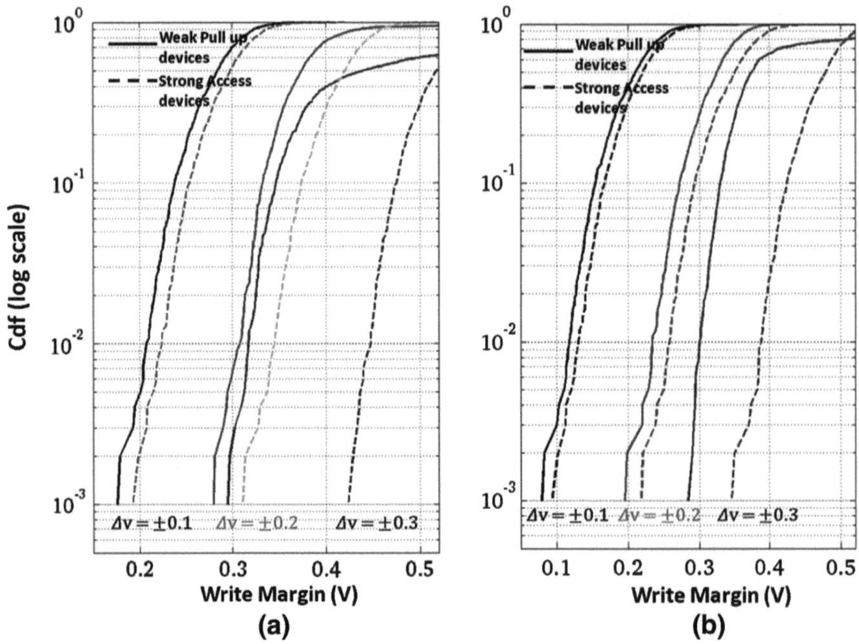

Fig. 3.14 Comparison of write-ability improvement: VDD = 0.8 V, Temp = 25°C and the worst process corner for write-ability (*slow nmos fast pmos*). Clearly strong NMOS pass transistor based techniques (WL Boost and Neg BL) outperforms the weak PMOS pull up transistor based techniques (VDD Low and VSS High) (a) 40 nm LP W = 150 nm, L = 40 nm, (b) 65 nm LP, W = 150 nm, L = 60 nm

case of word line boosting (WL Boost) and negative bit-line (Neg BL), the limiting factor is a nominal voltage of the technology (1.1 V for 40 nm LP and 1.2 V for 65 nm LP). In addition to this the value of an applied negative bit-line voltage is also limited by a potential risk of forward biasing the junction diodes (latch up). Similarly for the techniques relying on lowering the supply lines (VDD Low) and raising the VSS lines (VSS High) is limited by a potential risk of data retention failures for the un-accessed SRAM cells on the same modulated supply lines. Therefore, for the functional effectiveness analysis the value of the maximum applied bias at VDD = 0.8 V is kept at 0.3 V. The functional effectiveness of the WL Boost and the Neg BL technique is better compared to the VDD Low and VSS High (Fig. 3.14). For the higher values of the applied bias voltage the variability resilience of VDD Low is worst because of the reduced operating voltage levels (VDD = 0.5 and 0.55 V) with HVT SRAM cell transistors.

Figure 3.15, 3.16 shows the SNM read (cell stability) versus the applied bias voltage for the minimum sized HVT transistors based SRAM cell at VDD = 0.8 V. The functional effectiveness of lowering the VSS (Neg VSS) for improving the cell stability is maximum compared to the other assist techniques (Fig. 3.16). Due to the more pronounced impact of the Vt variations for the lower values of a bit-line

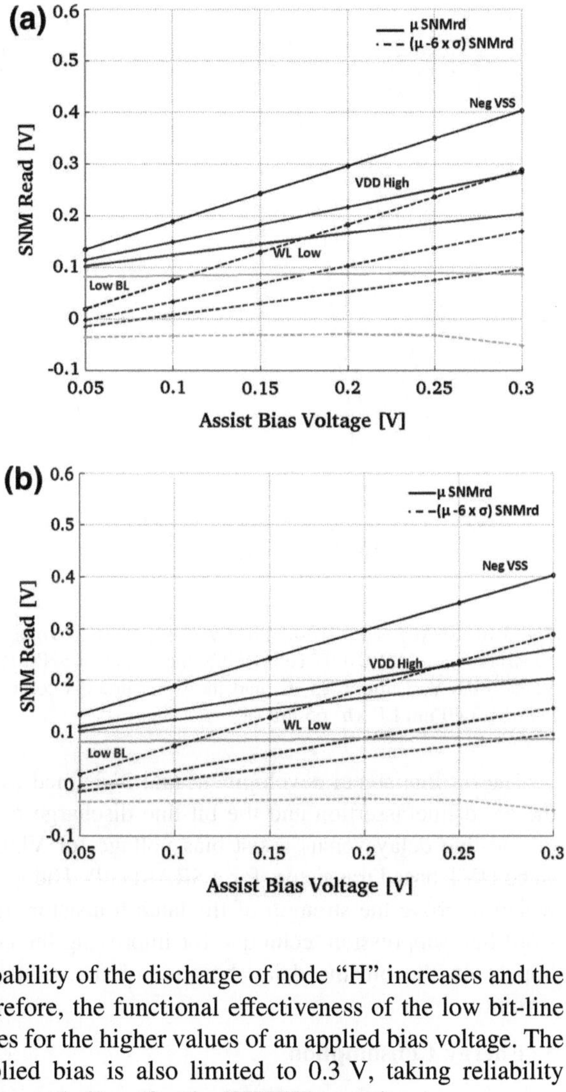

Fig. 3.15 Comparison of SNM read improvement: VDD = 0.8 V, Temp = 25 °C, and the worst process corner for cell stability (fast nmos slow pmos). Strong latch devices are better in improving cell stability (a) 40 nm LP W = 150 nm, L = 40 nm, (b) 65 nm LP, W = 150 nm, L = 60 nm

pre-charge voltage. The probability of the discharge of node "H" increases and the cell stability decreases. Therefore, the functional effectiveness of the low bit-line pre-charge technique degrades for the higher values of an applied bias voltage. The maximum value of the applied bias is also limited to 0.3 V, taking reliability concerns into account as discussed above for the WRITE operation.

2. **Performance**

Write delay is used as a metric for measuring the performance of the write assist techniques. Similarly for the read assist techniques the bit-line signal development rate is used as a metric for measuring the performance of the read assist techniques. Figure 3.17 shows the write delay (time difference between the word line assertion and the timing instance of data flip) versus the assist bias voltage. The voltage optimization increasing strength of a NMOS access transistor (WL Boost and Neg BL) results in the better performance.

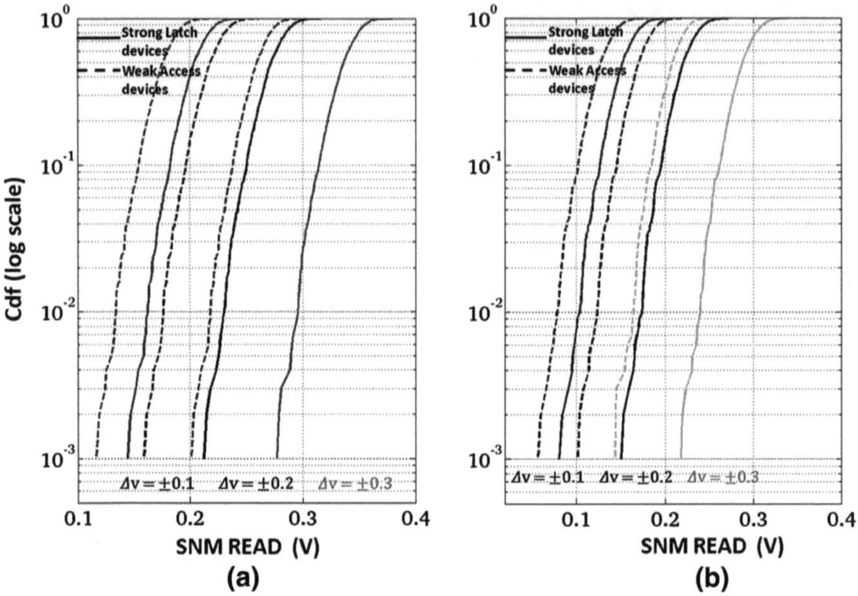

Fig. 3.16 SNM Read (cell stability) versus assist bias voltage Δv for the minimum sized HVT transistors based SRAM 6T cell. Higher the values of SNM Read means better the cell stability, VDD = 0.8 V, Temp = 25 °C, and the worst process corner for cell stability (*fast nmos slow pmos)* (**a**) 40 nm LP, (**b**) 65 nm LP

The bit line signal development rate is defined as the timing instance between the word line assertion and the bit-line discharge of 100 mV. Figure 3.18 shows the bit-line delay versus assist bias voltage for VDD = 0.8 V, for the minimum sized HVT based transistors for a SRAM cell. The voltage optimization techniques which improve the strength of the latch transistors reduce the bit-line delay. The word line suppression technique for improving the cell stability has a detrimental impact on the bit-line delay because of the reduction in the SRAM cell read current.

3. **Energy Consumption**

Energy overhead is an important criterion for the selection of the READ/WRITE assist technique. The energy overhead of the assist technique employed is also strongly dependent on its implementation. Therefore, for the fair comparison of energy consumption only the energy associated with the modulation (assist bias voltage application) is accounted. The energy consumption for the READ operation without using any assist technique is

$$E_{\mathrm{READ}} = E_{\mathrm{WL}} + E_{\mathrm{BL}} \tag{3.1}$$

E_{WL} Energy consumption associated with the word line (WL) assertion
E_{BL} Energy consumption associated with the bit-line discharge

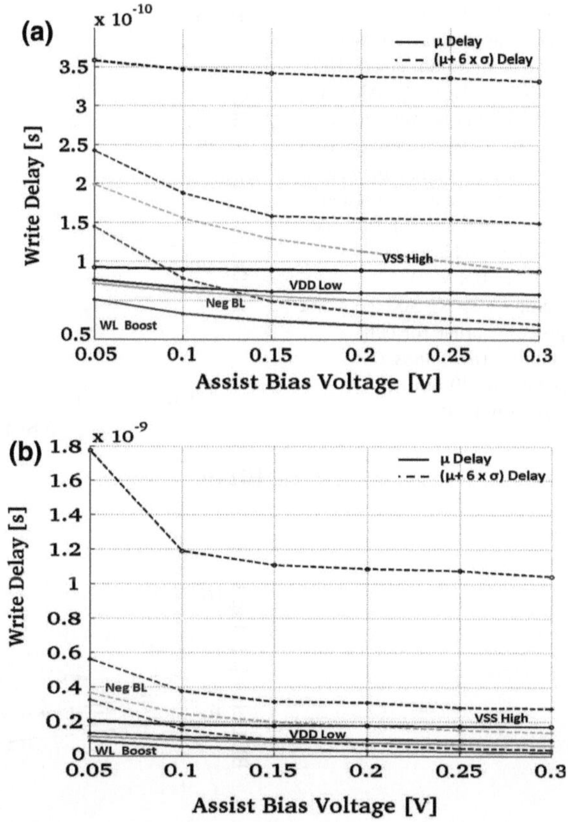

Fig. 3.17 Write delay versus assist bias voltage Δv for the minimum sized HVT transistors based SRAM 6T cell. The write delay is defined as the time difference between the word line assertion and the timing instance when the data flip occurs. VDD = 0.8 V, Temp = 25 °C, and the worst process corner for write-ability (*slow nmos fast pmos*) (**a**) 40 nm LP, (**b**) 65 nm LP

$$E_{\mathrm{WL}} = N_{\mathrm{wl}} \times C_{\mathrm{g,wl}} \times \mathrm{VDD}^2 \tag{3.2}$$

N_{wl} Number of cells in word (word length)
$C_{\mathrm{g,wl}}$ Word line gate capacitance of SRAM cell

$$E_{\mathrm{BL,bit}} = N_{\mathrm{v}} \times C_{\mathrm{d,bl}} \times \Delta V_{\mathrm{BL}} \times \mathrm{VDD} \tag{3.3}$$

N_{v} Number of cells in the vertical direction of SRAM array (column height)
$C_{\mathrm{d,bl}}$ Bit-line drain capacitance of SRAM cell
ΔV_{BL} The minimum bit-line discharge required for the sense amplifier

Substituting (3.3) and (3.2) in Eq. (3.1)

$$\begin{aligned} E_{\mathrm{READ}} &= N_{\mathrm{wl}} \times C_{\mathrm{g,wl}} \times \mathrm{VDD}^2 + N_{\mathrm{wl}} \times E_{\mathrm{BL,bit}} \\ &= N_{\mathrm{wl}} \times C_{\mathrm{g,wl}} \times \mathrm{VDD}^2 + N_{\mathrm{wl}} \times N_{\mathrm{v}} \times C_{\mathrm{d,bl}} \times \Delta V_{\mathrm{BL}} \times \mathrm{VDD} \end{aligned} \tag{3.4}$$

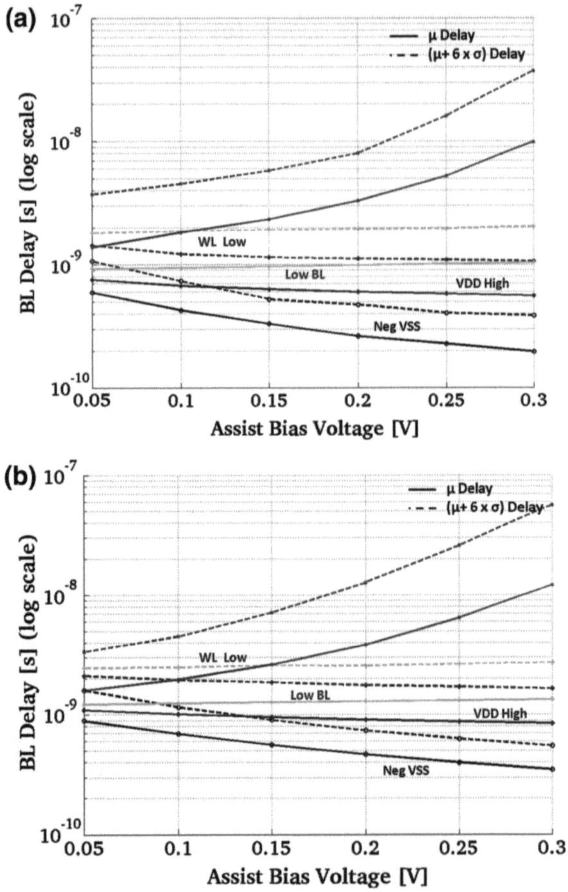

Fig. 3.18 BL (bit-line) delay versus assist bias voltage Δv for the minimum sized HVT transistors based SRAM 6T cell, column height 256 cells. The BL delay is defined as the time difference between the word line assertion and the development of 100 mV bit-line discharge. VDD = 0.8 V, Temp = 25 °C, and the worst process corner for cell stability (*fast nmos slow pmos*) (**a**) 40 nm LP, (**b**) 65 nm LP

Similarly, the energy consumption for the WRITE operation is defined as

$$E_{\mathrm{WRITE}} = E_{\mathrm{WL}} + E_{\mathrm{BLW}} + E_{\mathrm{WORD}} + E_{\Delta\mathrm{BL}} \tag{3.5}$$

E_{WL} Energy consumption associated with the word line (WL) assertion

E_{BLW} Energy consumption associated with the bit-lines switching from VDD to ground

E_{WORD} Energy associated with writing the selected cells on the word line

$E_{\Delta\mathrm{BL}}$ Energy due to the voltage change on the bit-lines along the asserted word line for the half selected cells

N_{h} Number of cells in the horizontal direction of SRAM array (number of bit-lines)

$$E_{\mathrm{BLW}} = N_{\mathrm{wl}} \times E_{\mathrm{BLW,bit}} = N_{\mathrm{wl}} \times N_{\mathrm{v}} \times C_{\mathrm{d,bl}} \times \mathrm{VDD}^2 \tag{3.6}$$

$$E_{\text{WORD}} = N_{\text{wl}} \times C_{\text{CELL}} \times \text{VDD}^2 \tag{3.7}$$

$$E_{\Delta\text{BL}} = (N_{\text{h}} - N_{\text{wl}}) \times C_{\text{CELL}} \times \Delta V_{\text{BL,hs\_cells}} \times \text{VDD} \tag{3.8}$$

Substituting Eqs. 3.2, 3.6, 3.7, and 3.8 in Eq. 3.5

$$\begin{aligned} E_{\text{WRITE}} = & N_{\text{wl}}\text{g, wl} \times \text{VDD}^2 + N_{\text{wl}} \times N_{\text{v}} \times C_{\text{d,bl}} \times \text{VDD}^2 \\ & + N_{\text{wl}} \times C_{\text{CELL}} \times \text{VDD}^2 + \\ & (N_{\text{h}} - N_{\text{wl}}) \times C_{\text{CELL}} \times_{\text{BL,hs\_cells}} \times \text{VDD} \end{aligned}$$

C_{CELL}	SRAM cell capacitance
$\Delta V_{\text{BL,hs\_cells}}$	The bit-line discharge caused by the half selected cells

The change in energy consumption associated with the assist technique is defined as

$$\begin{aligned} E_{\text{READ,assist}} &= E_{\text{READ},\Delta\text{v}} + E_{\text{assist}} \\ E_{\text{WRITE,assist}} &= E_{\text{WRITE},\Delta\text{v}} + E_{\text{assist}} \end{aligned}$$

$E_{\text{READ/WRITE},\Delta\text{v}}$	READ/WRITE energy consumption with the Δv (change in voltage, depending on the assist technique) applied
E_{assist}	Energy associated with the implementation overhead of the applied assist technique. For e.g. voltage reduction of the word line for improving cell stability using replica access transistors. Similarly, VDD switching for improving write-ability, power overhead is dependent on the SRAM array configuration (factors like column height, word length, SRAM cell layout, etc.)

1. **Word Line Modulation**

(a) Word line suppression during READ operation

$$E_{\text{READ},\Delta\text{v,WL}} = N_{\text{wl}} \times C_{\text{g,wl}} \times (\text{VDD} - \Delta v)^2 + E_{\text{BL,VDD}}$$

(b) Word line boosting during WRITE operation

$$\begin{aligned} E_{\text{WRITE}} = & N_{\text{wl}}\text{g, wl} \times (\text{VDD} - \Delta v)^2 + E_{\text{BLW}} \\ & + E_{\text{WORD}} + E_{\Delta\text{BL}} \\ & (N_{\text{h}} - N_{\text{wl}}) \times C_{\text{CELL}} \times \uparrow^{*} \Delta V_{\text{BL,hs\_cells}} \times \text{VDD} \end{aligned}$$

* the bit-line discharge caused by the half selected cells increases with word line boosting

2. **Cell Supply Modulation**

(a) VDD High during READ operation

$$E_{\mathrm{READ},\Delta\mathrm{v},\mathrm{WL}} = E_{\mathrm{WL}} + E_{\mathrm{BL,VDD}}$$

(b) VDD Low during WRITE operation

$$E_{\mathrm{WRITE}} = E_{\mathrm{WL}} + N_{\mathrm{wl}} \times C_{\mathrm{CELL}} \times (\mathrm{VDD} - \Delta v)^2 + E_{\mathrm{BLW}} + (N_{\mathrm{h}} - N_{\mathrm{wl}}) \times C_{\mathrm{CELL}} \times \downarrow^{*} \Delta V_{\mathrm{BL,hs\_cells}} \times (\mathrm{VDD} - \Delta v)$$

* the bit-line discharge caused by the half selected cells decreases with VDD lowering

3. **Bit-line Modulation**

(a) Low Bit-line pre-charge during READ operation

$$E_{\mathrm{READ},\Delta\mathrm{v},\mathrm{WL}} = E_{\mathrm{WL}} + N_{\mathrm{wl}} \times N_{\mathrm{v}} \times C_{\mathrm{d,bl}} \times \Delta V_{\mathrm{BL}} \times (\mathrm{VDD} - \Delta v)$$

(b) Negative bit-line during WRITE operation

$$E_{\mathrm{WRITE}} = E_{\mathrm{WL}} + N_{\mathrm{wl}} \times N_{\mathrm{v}} \times C_{\mathrm{d,bl}} \times (\mathrm{VDD} + \Delta v)^2 + E_{\mathrm{WORD}} + E_{\Delta\mathrm{BL}} \quad (3.9)$$

There are significant differences in the implementation of the assist techniques available in the literature. It becomes difficult to provide fair quantitative comparison. Table 3.1 provides a summary of the voltage optimization techniques. For the selection of a particular assist technique functionality requirement is the first and the most important criterion. The quantitative comparison of the performance analysis and the energy consumption associated with the modulation of the applied bias value also plays a crucial role in the selection process. Last, the qualitative analysis is always beneficial in estimating the implementation cost of the assist technique selected.

3.5 Hybrid Voltage Optimization Techniques

This section covers another category of voltage optimization based assist techniques which combines modulating 2 or more terminal voltages. This enables the circuits to have necessary adaptability for different voltage ranges. These techniques target to offer better variability resilience for the same value of the assist bias voltage applied. Crosshairs SRAM (Chen et al. 2010) provides an independent tuning of VDD and GND of each bit cell inverter. The performance improvement with Crosshairs SRAM is much higher than VDD lowering and VSS raising when applied individually. Configurable WR assist technique (Sinangil et al. 2009)

Table 3.1 Summary of voltage optimization techniques

	Technology	Performance	Capacity	Min VDD	Area overhead (%)	Read assist technique	Write assist technique
(Yamaoka et al. 2004)	130 nm	300 MHz @ 1.2 V	1 Mb	0.8 V	–	VDD high	VSS high
(Zhang et al. 2006)	65 nm	3 GHz@ 1.1 V	70 Mb	0.9 V	–	VDD High	VDD low
(Yamaoka et al. 2006)	90 nm	450 MHz@ 1.2 V	512 Kb	0.8 V		–	VDD floating
(Khellah et al. 2006)	90 nin	–	–	0.7 V	(4–8)	Pulsed BL	Pulsed WL
(Ohbayashi et al. 2007)	65 nmLP	–	512 Kb	1.2 V	<2	WL suppression (replica access Tr.)	VDD lowering
(Yabuuchi et al. 2007)	45 nmLP	–	1 Mb	1.0 V	<10	WL Suppression (gate controller RAT)	VDD Lowering (capacitive write assist)
(Nii et al. 2008)	45 nm LP	278 MHz @ 1.1 V	1.5 Mb (DP SRAM)	0.7 V	4	WL suppression (gate controller RAT)	Negative BL (cap boost circuit)
(Hirabayashi et al. 2009)	40 nm	417 MHz @ 1.0 V	2 Mb	0.7 V	(2–4)	WL suppression (LPWD)	WL boosting + Low VDD (LPWD + DASC)
(Fujimura et al. 2010)	32 nm H–K	–	1.25 Mb	0.5 V	(2–3)	WL Suppression (LPWD)	Negative BL (configurable)
(Raychowdhury et al. 2010)	45 nm	4 GHz @ 0.8 V	16 KB (DP SRAM)	0.45 V	(8–17)	Read WL boosting (charge pump and LS)	WL boosting (charge pump and LS)
(Sharma et al. 2010)	90 nm	80 MHz@ 0.8 V	128 Kb	0.8 V	–	VDD/2 short local bit-lines + local sense amplifier	Selective VDD lowering

activates one of the three different schemes (VDD fixed, VDD floating, and VDD collapsing) depending on the operating VDD supplies. MNBL technique (Sharma et al. 2011) resolve the issues associated with the conventional negative bit-line technique (latch up issue) as it does not require a negative bit-line voltage. The CDVSS (Sharma et al. 2012) offers 2 assist techniques at the cost of 1 applied assist bias and also the performance of CDVSS (Sharma et al. 2012) bias is higher compared to the Crosshair SRAM (Chen et al. 2010).

3.5.1 Crosshairs SRAM—Separately Tuning VDD and GND Supplies of SRAM Cells

The crosshairs SRAM technique mitigates the process variation by tuning VDD and VSS of each bitcell inverter independently from its cross-coupled counterpart (Chen et al. 2010). Each SRAM bit cell has connections to the left and the right vertical power rails (VDDL and VDDR) and the horizontal ground rails (VSSL and VSSR). VDDL and VDDR rails are shared for a matrix column and VSSR and VSSL are shared for adjacent rows. The write failures are detected by BIST which then tunes VDD and VSS of each bitcell inverter with respect to its cross-coupled counterpart. VDD tuning is done by the PMOS headers by connecting VDDR and VDDL to one of two global power supplies (VDD and VDD $-$ ΔV). Similarly VSS is tuned by the NMOS footers by connecting VSSR and VSSL to either VSS + ΔV or VSS. Figure 3.19 shows the implementation of Crosshairs SRAM. During WRITE operation PUL is weakend by connecting VDDL to VDD $-$ ΔV and VSSR to VSS + ΔV. It reduces the strength of PUL and PDR and the write failures are avoided. Similarly during the READ operation transistor strength, holding the state of cell is increased by connecting VDDR to VDD + ΔV and VSSL to VSS $-$ ΔV and the read upset failures are avoided. Figure 3.20 shows performance gain with Crosshairs SRAM.

3.5.2 Configurable Write Assist: Compatibility with a Dynamic Voltage Scaling

Sinangil et al. (2009) proposes a reconfigurable write assist scheme enabling write-ability of the cell over the entire voltage range (1.2 V–0.25 V) of dynamic voltage scaling compatible designs. It does not require any external power supplies or capacitive circuits to enable VDD lowering. Figure 3.21 shows implementation of a configurable write assist technique. VDD Cell Line is a row-wise virtual supply node. It is connected to all memory cells on the same row which will be accessed during WRITE. Depending on the operating voltage range, a reconfigurable write assist circuit activates one of the three different schemes. At super threshold voltage levels, memory cells have enough write margins to operate correctly so VDD Cell

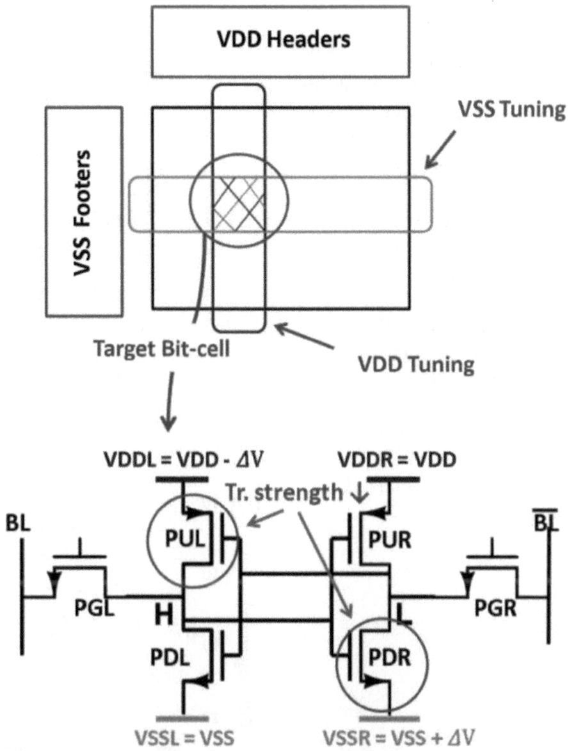

Fig. 3.19 Crosshair implementation (Chen et al. 2010)—separate tuning of the VDD and VSS supplies of each inverter within a bitcell. VDD within a column and VSS within a row are tuned to correct the functionality of the target bitcell. Headers connect each VDD column (VDDL, VDDR) to two global supplies. Footers connect VSS row (VSSL, VSSR) to two global grounds

Line is kept at VDD. This avoids unnecessary switching in the control and row circuitry and avoids power consumption. The floating VDD Cell Line for the scaled supply voltages improves the write-ability. The VDD Cell Line voltage droop helps the access transistors to overpower the PMOS transistors of the accessed cells more easily and the write failures are prevented. At sub threshold voltages the VDD Cell line is actively pulled down to VSS in order to prevent write failures. As the floating VDD Cell Line node requires significantly larger amount of time to droop. This is how the reconfigurable write assist technique is implemented in order to solve the write-ability issues for different operating voltage levels.

3.5.3 MNBL Technique: Sequential Voltage Optimization

Write-ability of SRAM is improved by reducing the strength of a pull up PMOS transistor or by increasing the strength of the NMOS pass transistor (Sharma et al. 2011). The amount of assist bias voltage applied for either increasing the strength of the pass transistor or for reducing the strength of a pull up transistor is limited by the half select stability issues and the data retention issues. The half select stability issues

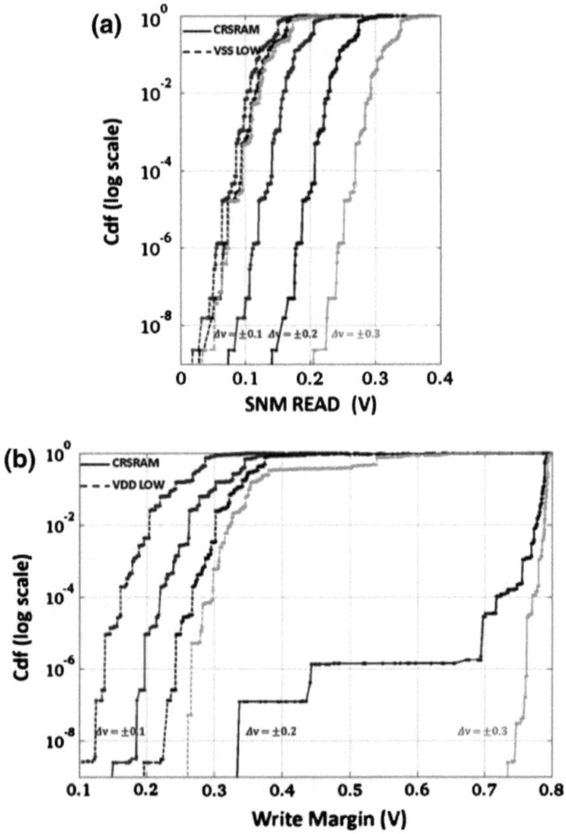

Fig. 3.20 Comparison of cell stability and write-ability improvement: VDD = 0.8 V, Temp = 25 °C. The SNM read improvement for the values of applied bias voltage is much higher with Crosshair SRAM compared to the Neg VSS technique. Similarly for the WM improvement Crosshair SRAM offers higher margin compared to the single VDD lowering technique (**a**) SNM READ 40 nm (*worst corner: fast nmos slow pmos*) (**b**) WM 40 nm (*worst corner: slow nmos fast pmos*)

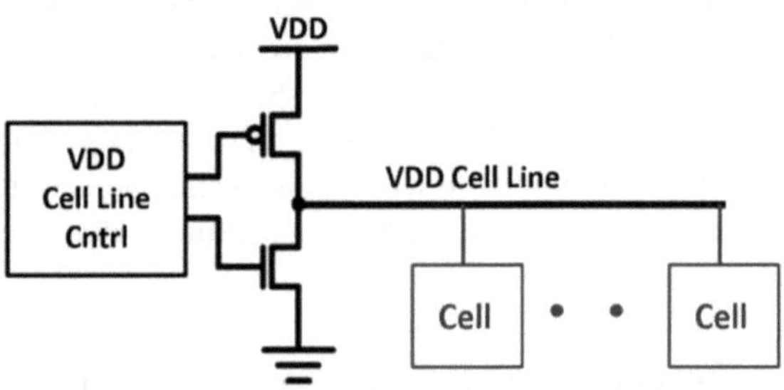

VDD Cell Line levels for different operating ranges

Super Threshold Voltages	Near Threshold Voltages	Sub Threshold Voltages
VDD	Floating	Collapsed to VSS

Fig. 3.21 Configurable Write Assist technique (Sinangil et al. 2009). Three different write assist schemes are used in the U-DVS SRAM. For super threshold voltages (1.2–0.8)V VDD Cell Line is connected to VDD. For reduced supply voltages above/near threshold (0.8–0.5)V VDD Cell Line is kept floating and for sub threshold (0.5–0.25)V the VDD Cell Line is pulled down to 0 V. Implementation 64 Kb SRAM macro in 65 nm

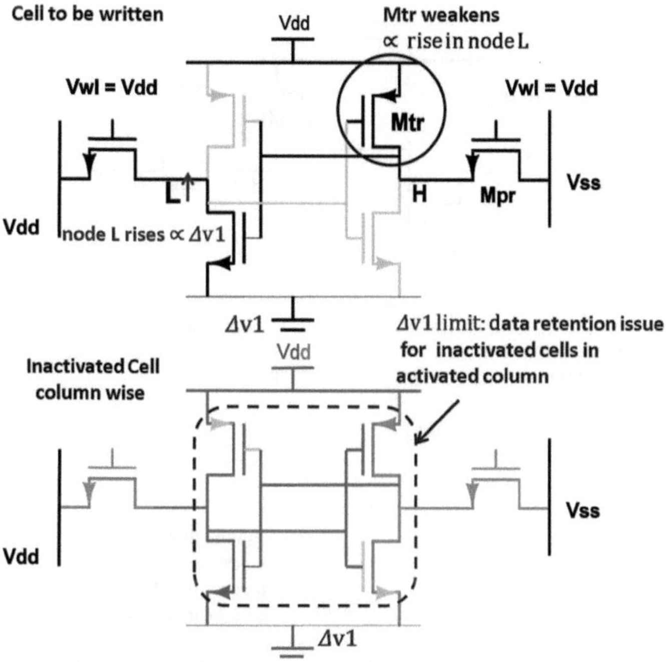

Fig. 3.22 VSS biasing: Column-wise VSS raising step1

and the data retention issues are problematic especially at the scaled voltage levels for the unselected cells in the activated column and activated row of the memory matrix. The sequential voltage optimization technique (Sharma et al. 2011) remedies the above-mentioned issue and ensures a high write-ability for the scaled voltage levels.

3.5.3.1 Concept

1. VSS (GND) of the activated columns of memory matrix is raised by Δv1. It is desirable to do the voltage optimization of VSS in the first go because of the latency associated in changing the voltage of highly capacitive VSS supply lines for the activated columns. Increasing the voltage of VSS by Δv1 weakens the strength of Mtr (aids in improving write-ability) (Fig. 3.22). The reduction in the strength of Mtr is proportional to the value of Δv1 applied on the VSS supply lines. The Δv1 is limited by the data retention issues of inactivated cells on the activated column. The operating margin of the inactivated cells reduces with the amount equal to Δv1 applied.
2. Then the WL is modulated by Δv2, which increases the strength of the Mpr (aids in improving write-ability). Together Δv1 + Δv2 applied results in the higher values of write margin. Higher the value of Δv2 higher will be the write margin. But the value of WL modulation Δv2 is limited by the half select condition of the inactivated cells in an activated column (Fig. 3.23).

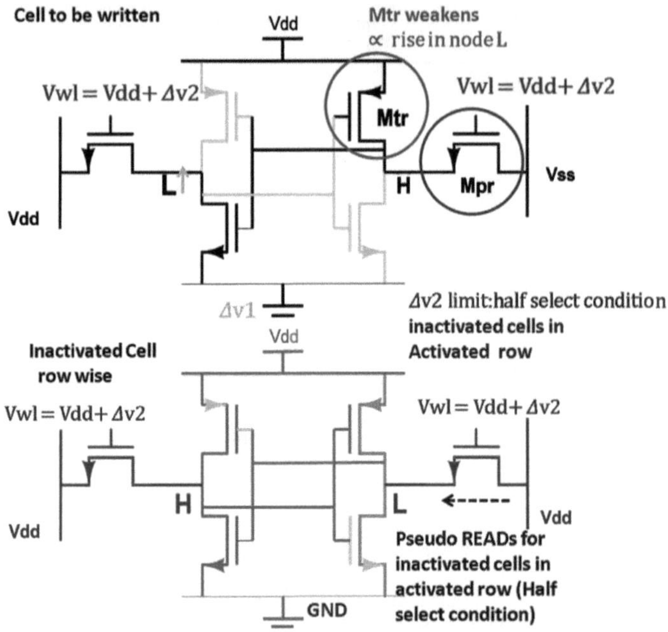

Fig. 3.23 Vwl biasing: WL driver supply line of an activated row is raised

3. Figure 3.24 in the final step VDD supply line (row wise and column wise) are modulated to a higher value Δv3. The modulation of the VDD supply lines reduce the probability of data flips arising due to the half select condition of the inactivated cells in an activated row. However, application of Δv3 reduces the write-ability of the asserted cells by increasing the strength of Mtr. The application of Δv1 reduces the strength of Mtr (increase write-ability) and application of Δv3 reduces the strength of Mtr (decreases write-ability and increases inactivated cells stability).

The order of step 2 and step 3 can be interchanged depending on the scenario (if cell stability is important or the write-ability is more important). It is also possible to share the VDD supply lines of SRAM cells and WL drivers in that case step 2 and step 3 happens simultaneously and the value of Δv2 = Δv3.

Figure 3.25, shows the distribution of write margin for different values of Δv, VDD = 0.55 V, 65 nm LP technology. The higher values of assist bias targeting the write-ability improvement (Δv1, Δv2) compared to the assist bias targeting the data retention issue and the half select condition (Δv3) yields higher improvement in the write margin compared to the situation when Δv3 > (Δv1, Δv2).

Figure 3.26, shows the distribution of SNM read for different values of Δv, VDD = 0.55 V, 65 nm LP technology. The higher values of assist bias targeting the write-ability improvement (Δv1, Δv2) compared to the assist bias targeting the

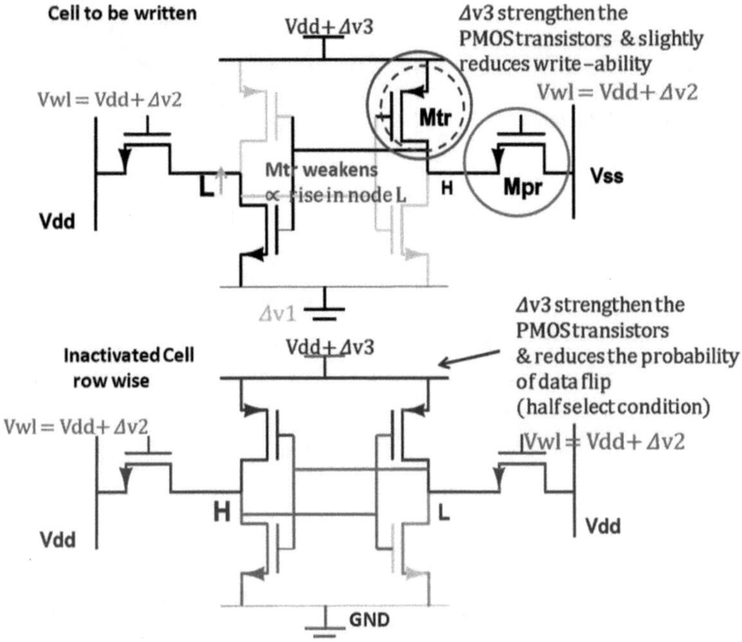

Fig. 3.24 VDD biasing: VDD supply line of an activated row is raised

Fig. 3.25 Distribution of write margin for different values of Δv, VDD = 0.55 V, 65 nm LP technology

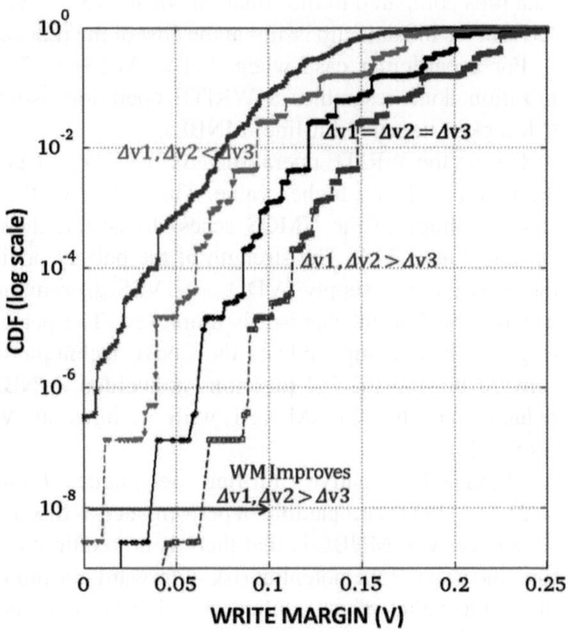

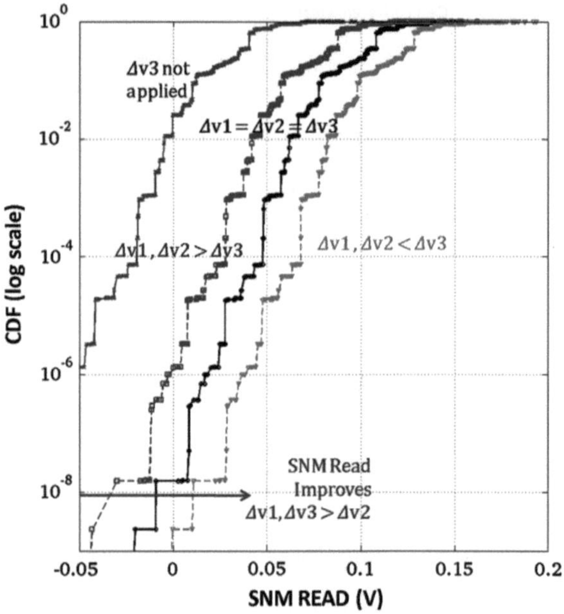

Fig. 3.26 Distribution of SNM Read (cell stability) for different values of Δv for inactivated cells in activated row, VDD = 0.55 V, 65 nm LP technology

data retention issue and the half select condition (Δv3) results in a higher risk of data flips compared to the situation when Δv3 > (Δv1, Δv2). The higher values of v3 resolves the data flip issues at the cost of the reduced write margin improvement.

For a particular case when Δv1 = Δv2 = Δv3 = Δv, then the voltage optimization done resembles a WRITE operation using negative bit-line technique (Mimicked negative bit-line, MNBL).

During the WRITE operation, (Δv3) VDD_{cell}, (Δv1) VSS_{cell}, and (Δv2) VDD_{wl} are modulated to a higher value. This enhances the gate-to-source and drain-to-source voltage of the NMOS access transistor, thereby increasing the discharge current (Fig. 3.27a). The strength of the pull up devices does not change because the effective cell supply ($VDD_{cell} - VSS_{cell}$) remains the same. This is how the negative bit-line mechanism is mimicked. The performance of MNBL technique (Fig. 3.27b) is comparable to the CNBL technique and also the potential risk of forward biasing the PN junctions is avoided. MNBL technique results in $10^3\times$ reduction in the SRAM cell write failures at VDD = 0.66 V and −20 °C (Fig. 3.28).

Mimicked negative bit-line technique (condition when Δv1 = Δv2 = Δv3 = Δv) is comparable in performance as discussed earlier. But the advantage associated with MNBL is that there is no requirement of negative bit-line. Therefore, the issue of the potential risk of forward biasing PN Junctions (latch up) which limits the value of an applied negative boost associated with the conventional negative bit-line technique is resolved with a new innovative MNBL technique.

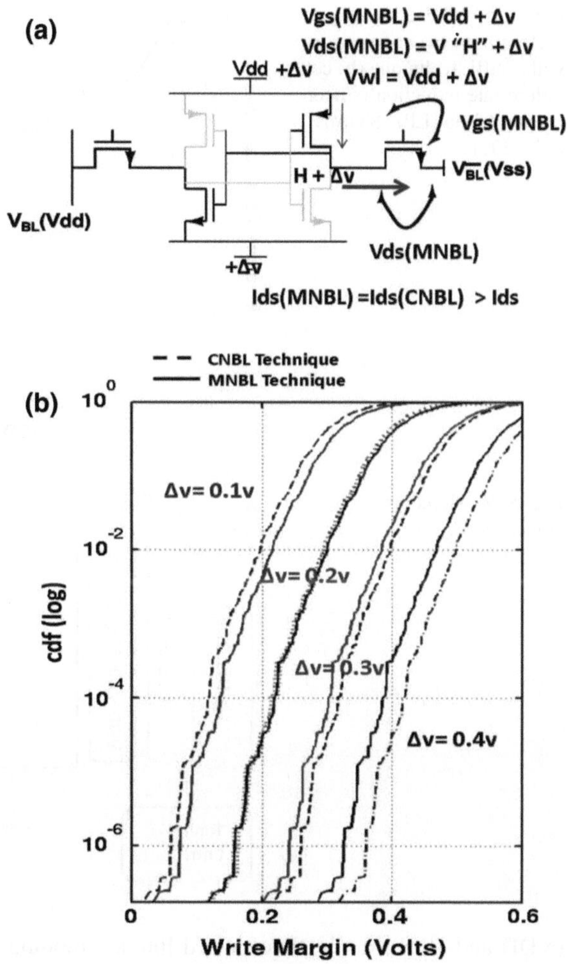

Fig. 3.27 MNBL technique: (**a**) Concept (**b**) Comparison with the conventional negative bit-line (CNBL) technique

3.5.3.2 Implementation

Sharma et al. (2011) outlines only the concept of MNBL technique. The generation of tuning voltages Δv and the modulation of VDD and VSS are implemented off-chip but can be implemented on—chip as a part of future work. Figure 3.29 shows the memory floorplan for the MNBL implementation. At the column level 2 power supplies are routed {VDD + Δv3 and VSS + Δv1}. For the row level also 2 power supplies are routed {VDD + Δv3 and VDDwl + Δv2}. The voltage optimization for MNBL is performed for the activated column and the activated row. The word decodes logic and the tuning circuitry dynamically adjusts VDD_SRAM and VSS_SRAM (Fig. 3.30).

During WRITE operation VDD_SRAM, VDD_WL, and VSS_SRAM are adjusted to a higher value viz. VDD_high/VSS_high with the result cell supply

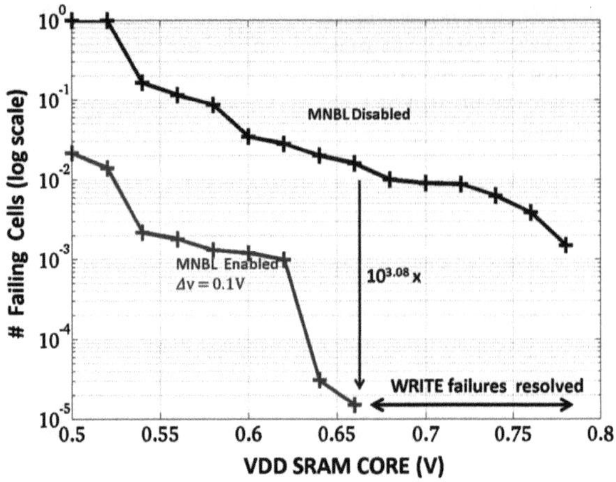

Fig. 3.28 MNBL technique performance (**a**) comparison with CNBL technique (**b**) cell failure rate reduction (silicon results, 65 nm LP) (Sharma et al. 2011)

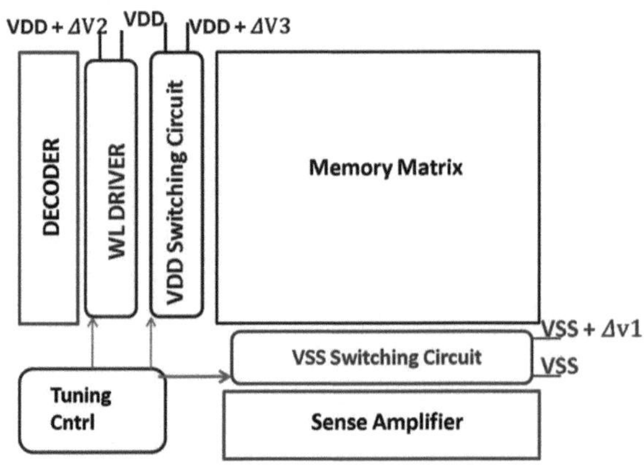

Fig. 3.29 Memory Organization

(VDD and VSS), the VDD of Word line are modulated to a higher value thereby realizing Mimicked Negative Bit-line technique. During the READ operation regular power supply values VDD/VSS are used.

3.5.4 Compounded Differential VSS (CDVSS) Bias Technique

The Crosshair SRAM (Chen et al. 2010) tunes VDD and GND of each bit cell inverter with respect to its cross-coupled counterpart independently in order to fix write failures (Sharma et al. 2012). The simultaneous tuning of both VDD and GND rails not only increases the design complexity but also results in an increased area and power consumption overhead. This issue is remedied by biasing only GND rails in such a manner that it automatically compounds into 2 voltage optimizations (VSS biasing and Neg BL) as proposed in (Sharma et al. 2012).

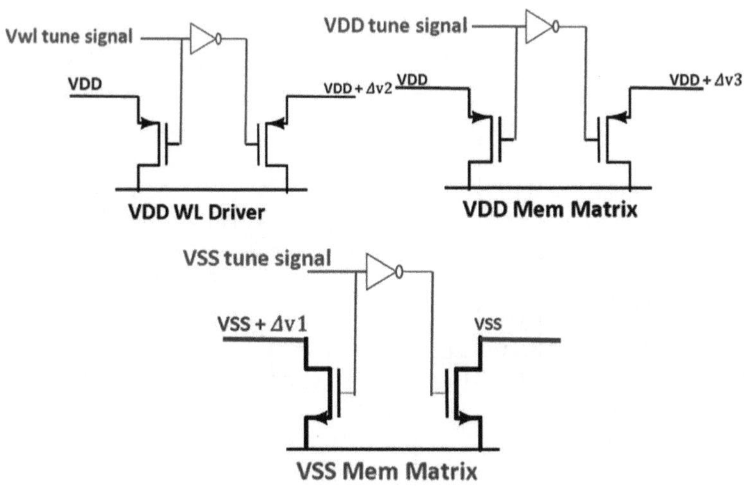

Fig. 3.30 Switching Circuit: tune signal (Vwl, VDD, and VSS) are generated from the timing circuit. Initial condition before using voltage optimization VDD, VSS, and Vwl tune signal is "L" and VDD driver and memory matrix are powered from VDD and ground of memory matrix is at VSS

The differential VSS bias enabled non strobed local write receiver (Sharma et al. 2012) provides a combined solution for realizing variability resilient and low energy WRITE operation. The detailed local architecture and low energy WRITE operation will be covered in the next chapters. The write failures can be either due to sensing failure of the local write receiver (SRAM 6T cell type structure) or because of the degraded write-ability of the accessed SRAM cell. Due to the process variations, if MUP transistor of a local write receiver becomes strong and MUPbar becomes weak (Fig. 3.31a) due to process variations, then the risk of sensing failure increases (for writing "0"). Similarly, if MUP of SRAM cell becomes stronger and Mpass becomes weaker (Fig. 3.31b) then the discharge of the node H becomes more difficult and the write-ability of the SRAM cell decreases.

The application of differential VSS bias connects VSSL to + Δv makes MUP transistor weak. The connection of VSSR to −Δv increases the strength of MUPbar. The mismatch offset is reduced and the sensing failure is avoided. Differential VSS biasing of 0.1 V reduces the sigma Voffset by 25 %, based on the importance sampling simulations at VDD 0.55 V. The differential VSS bias bipartite into two writes assist techniques viz. the selective VSS raising and the negative bit-line mechanism for the accessed SRAM cell (Fig. 3.31b). The positive VSS bias applied weakens MUP of "H" side of the SRAM cell thereby improving write-ability of the accessed SRAM cell. The negative VSS bias applied on the complement GND signal have 2 advantages: first it makes the rise time faster during the WRITE operation thereby improving the write access time, −0.1 V of differential VSS bias results in 24 % improvement for the slow NMOS and slow

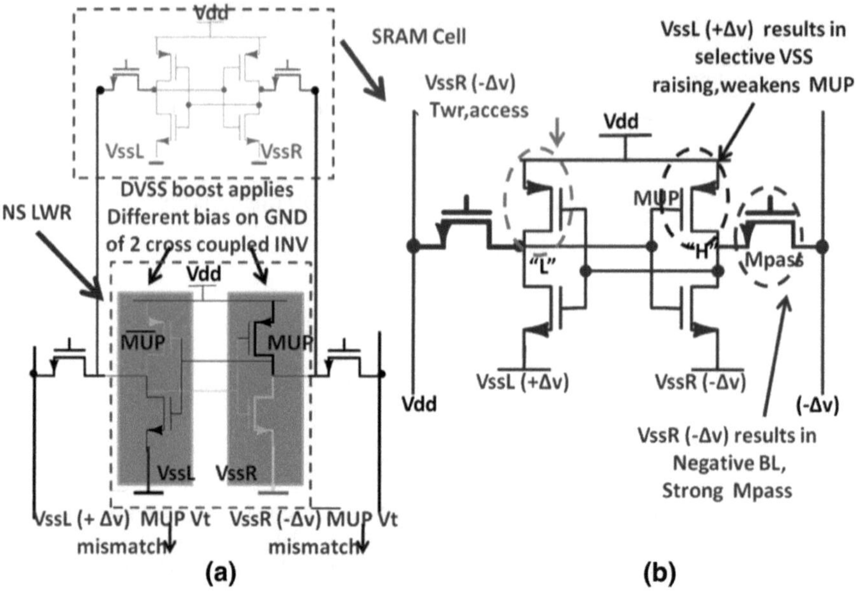

Fig. 3.31 Differential VSS bias cancels mismatch offset of NS-LWR and improves write margin for SRAM cell. Unlike the Crosshair SRAM applied voltage optimization, VDD, and GND). The application of differential VSS bias bipartite into VSS raising and negative bit-line technique (2 voltage optimizations at the cost of only VSS bias applied).This also rules out the possibility for applying higher values for the assist bias voltage, so there is a less risk of reliability concerns and the latch up with the negative bit-line technique

PMOS process corner. Second it pulls the bit-line below GND level ($-\Delta$v) and generates the negative bit-line for the accessed SRAM cell without any extra added cost. The selective VSS raising and negative bit-line mechanism increases the SRAM cell write-ability (Fig. 3.32) and the probability of write failure for the worst corner (slow NMOS and fast PMOS) by a factor of $10^3\times$ at the scaled VDD levels (VDD = 0.55 V).

The write margin improvement with the CDVSS bias technique for VDD = 0.55 V is comparable to the Crosshair SRAM operating at 0.6 V for the application of the same applied bias voltage of $\pm$ 0.1 V (Fig. 3.33).

3.6 Summary

This chapter provides an in-depth analysis of adaptive voltage optimization techniques for realizing low VDD SRAM. Sections 3.2 and 3.3 discuss the implementation of the state-of-the-art adaptive voltage optimization techniques for improving the variability resilience of the SRAM cell. Table 3.1, provides a summary for the various adaptive voltage optimization techniques. Section 3.4,

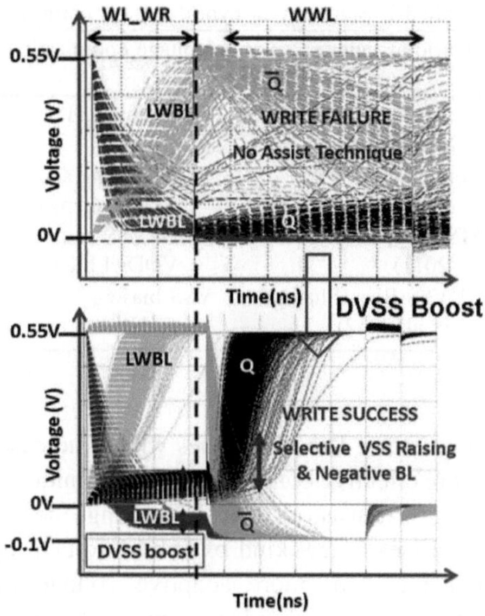

Fig. 3.32 Variability Resilience: SRAM cell write-ability improvement. VDD = 0.55 V, worst process corner (*slow nmos fast pmos*). The probability of write failure is reduced by the factor of $10^3\times$ at the scaled VDD levels (VDD = 0.55 V)

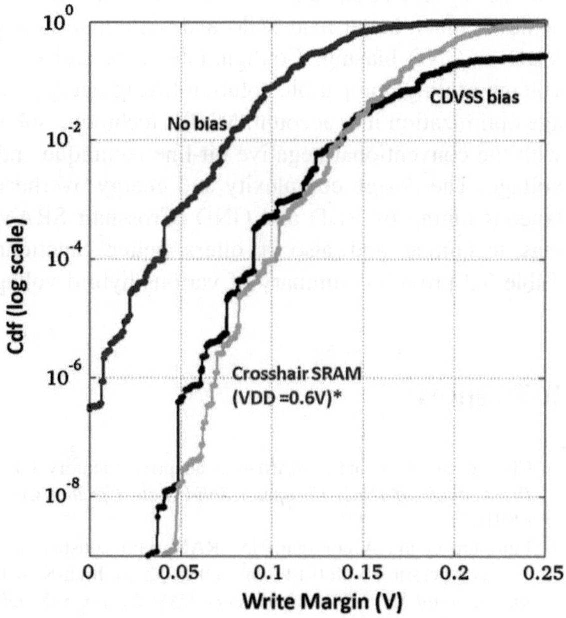

Fig. 3.33 Comparison of write margin of the HVT transistor based SRAM cell at VDD = 0.55 V, ± 0.1 V applied bias, and worst process corner (*slow nmos fast pmos*). Due to the data retention problem with Crosshair SRAM the value of VDD taken is 0.6 V. The functional effectiveness of CDVSS bias technique for VDD = 0.55 V is comparable to the Crosshair SRAM operating at 0.6 V for the same applied bias of ± 0.1 V

evaluates several adaptive voltage optimization techniques and compares the functional effectiveness, performance, and the energy consumption for the applied bias voltage value of the given assist technique. The voltage optimization which increases the strength of the NMOS access transistors of the SRAM cell are better compared to the techniques which target reducing the strength of the latch

Table 3.2 Summary of hybrid voltage optimization based techniques

Hybrid technique	Voltage optimization	WR margin improvement	Implementation overhead
Crosshairs SRAM (Chen et al. 2010)	VDD Low and VSS high	Very high	Voltage regulators and power switches
Configurable WR (Sinangil et al. 2009)	VDD floating and VDD collapsing	Medium	Power switches
MNBL (Sharma et al. 2011)	VDD, VSS and VDDwl high	High	Voltage regulators and power switches
CDVSS Bias (Sharma et al. 2012)	VSS biasing ($+\Delta$v, $-\Delta$v)	Highest	Voltage regulators and power switches

transistors. The techniques which increase the strength of the latch transistors are more effective compared to the techniques which reduce the strength of the NMOS access transistors, for improving the read SNM (cell stability). Section 3.5, discusses a new kind of hybrid voltage optimization techniques which combines existing 2 or more adaptive voltage optimization techniques to yield better performance and solves some of the key issues related to the existing adaptive voltage optimization techniques. Crosshair SRAM (VDD and GND tuning) achieves much better read SNM and write margin improvement compared to only VDD or GND biasing. Configurable write assist technique provides a dynamic voltage scaling compatible solution taking energy overhead associated with voltage optimization into account. MNBL technique solves the issue (latch up) related with the conventional negative bit-line technique and enables a higher assist bias voltage. The design complexity and energy overhead associated with the simultaneous tuning of VDD and GND (Crosshair SRAM) is remedied with CDVSS bias technique and also it offers better functional effectiveness (Fig. 3.33). Table 3.2 provides summary of various hybrid voltage optimization techniques.

References

G. Chen et al., Crosshars SRAM—an adaptive memory for mitigating parametric failures. in *Proceedings of IEEE European Solid State Circuits Conference (ESSCIRC)*, pp. 366–369 (2010)

Y. Fujimura et al., A configurable SRAM with constant-negative-level write buffer for low-voltage operation with 0.149um^2 cell in 32 nm high-K metal-gate CMOS. in *Proceedings of International Solid State Conference (ISSCC)*, pp. 348–350, Feb 2010

O. Hirabayashi et al., A process-variation-tolerant dual-power-supply SRAM with 0.179um2 Cell in 40nm CMOS using level-programmable wordline driver. in *Proceedings of International Solid State Conference (ISSCC)*, pp. 458–459, Feb 2009

M. Khellah et al., Word line and bitline pulsing schemes for improving SRAM cell stability in low-Vcc 65 nm CMOS designs. Symposium on VLSI circuits digest of technical papers, pp. 9–10 (2006)

K. Nii et al., A 45-nm single-port and dual-port SRAM family with robust read/write stabilizing circuitry under DVFS environment. Symposium on VLSI circuits digest of technical papers, pp. 212–213 (2008)

S. Ohbayashi et al., A 65-nm SoC embedded 6T-SRAM designed for manufacturability with read and write operation stabilizing circuits. IEEE J. Solid State Circ. **42**(4), 820–829 (2007)

A. Raychowdhury et al., PVT-and-aging adaptive wordline boosting for 8T SRAM power reduction. in *Proceedings of International Solid State Conference (ISSCC)*, pp. 352–353 (2010)

M. Sinangil et al., Reconfigurable 8T ultra-dynamic voltage scalable (U-DVS) SRAM in 65 nm CMOS. IEEE J. Solid-State Circ. **44**(11), 3163–3173 (2009)

M. Sinangil et al., A 28 nm high density 6T SRAM with optimized peripheral- assist circuits for operation down to 0.6 V. in *Proceedings of International Solid State Conference (ISSCC)*, pp. 260–261, Feb 2011

V. Sharma et al., A 4.4pJ/Access 80 MHz, 128 kbit variability resilient SRAM with multi-sized sense amplifier redundancy. IEEE J. Solid State Circ. **46**(10) (2010)

V. Sharma et al., 8T SRAM with mimicked negative bit-lines and charge limited sequential sense amplifier for wireless sensor nodes. in *Proceedings of IEEE European Solid State Circuits Conference (ESSCIRC)*, pp. 531–534 (2011)

V. Sharma et al., Ultra low power litho friendly local assist circuitry for variability resilient 8T SRAM cell, Design automation and test in Europe (DATE), Dresden, 11–17 March 2012

M. Yabuuchi et al., A 45 nm low-standby-power embedded sram with improved immunity against process and temperature variations. in *Proceeding of ISSCC*, pp. 326–327, 606 (2007)

M. Yabuuchi et al., A 45NM 0.6 V cross-point 8T SRAM with negative read/write assist. Symposium on VLSI circuits digest of technical papers, pp. 158–159, (2009)

M. Yamaoka et al., A 300 MHz 25uA/Mb leakage on-chip SRAM module featuring process-variation immunity and low-leakage-active mode for mobile-phone application processor. in *Proceedings of International Solid State Conference (ISSCC)*, Section 27.2 (2004)

M. Yamaoka et al., 90-nm process-variation adaptive embedded SRAM modules with power-line-floating write technique. IEEE J. Solid State Circ. **41**(3), 705–711 (2006)

K. Zhang et al., A 3-GHz 70-Mb SRAM in 65-nm CMOS technology with integrated column-based dynamic power supply. IEEE J. Solid State Circ. **41**(1), 146–151 (2006)

Chapter 4
Circuit Techniques to Assist SRAM Cell: Local Assist Circuitry

4.1 Introduction

SRAM cell design is a critical feature in achieving technology scaling benefits for SOC designs. The reduced voltage level and the utilization of low-power (LP) CMOS technologies are required for the low leakage SRAM cell design. The reduction in VDD and the increased threshold voltage aggravates variability for the SRAM cell design. This result in degradation of Iread, read static noise margin (SNM), and write-ability (WM) of SRAM cell as discussed in Chap. 2. The design optimizations done in improving one parameter often ends up in worsening the other. Read SNM (functionality) is of utmost concern and SRAM design techniques to improve the read SNM come at the expense of detrimental impact on Iread. Therefore, conventional SRAM 6T cell design is a highly constrained area-stability-power-performance trade off design effort. Local assist circuit techniques with hierarchical bit-lines are becoming increasingly necessary to maintain the SRAM cell functionality and to achieve performance target at the cost of a minimal area increase. The use of local assist circuits alleviates the complex design trade off effort of SRAM cell design. This chapter discusses various circuit assist techniques to alleviate the complex design trade off of SRAM cell design. The different assist techniques discussed are as follows

1. Hierarchical divided bit-lines (Kar 98 and Yang and Kim 2005): reduction of effective bit-line capacitance for accessed SRAM cell. Option for local write receiver as proposed by Yang and Kim (2005) reduces the WRITE energy consumption.

2. Hierarchical divided bit-lines with local assist circuitry (Kawasumi et al. 2008; Ishikura et al. 2008; Chang et al. 2008 and Cosemans et al. 2007): addition of an upsized low Vt read buffer as a local assist circuit accelerates the bit-line discharge rate and achieves high performance. The low swing pre-charged global bit-lines as proposed by Cosemans et al. (2007) further increases the variability resilience.

V. Sharma et al., *SRAM Design for Wireless Sensor Networks*,
Analog Circuits and Signal Processing, DOI: 10.1007/978-1-4614-4039-0_4,

3.WRITE after READ based assist circuitry (Pilo et al. 2007; Cosemans et al. 2009; Yoshimoto et al. 2011) enables DRAM-type sensing operation by rewriting the cell content after every READ operation.

4. Low swing bit-line hierarchy: enhanced SRAM cell stability (Sharma et.al. 2010): Reduced pre-charge voltage for short local bit-lines further enhances SRAM cell stability. Pseudo 8T sensing enabled local assist circuitry (Sharma et.al. 2010) promises an energy efficient READ/WRITE operation for low performance applications (<100 MHz). Hierarchical buffered segmented bit-lines are an alternative enhanced SRAM cell stability based local architecture for high performance SRAMs.

5. High bit density based hierarchy (Kushida et al. 2009; Sharma et al. 2012): alleviates the area overhead issue associated with the use of local assist circuit.

4.2 Hierarchical Divided Bit-Lines

The bit-lines are divided into short local bit-lines (LBL) that connect through pass transistors to global bit-lines (GBL). READ operation consists of pre-charging the local and global bit-lines (Kar 98 and Yang and Kim 2005).

The word line is asserted before activating the pass transistor between the local and global bit-lines. The local bit-lines are completely discharged to ground and the word line is disabled. Then the activation of the pass transistor between LBL and GBL initiates the capacitive charge distribution between LBL and GBL. This is how a well-controlled small voltage swing on GBL is generated. The hierarchical divided bit-line architecture benefits from the dynamic read stability, due to the reduced effective bit-line capacitance with the hierarchical bit-lines. The risk of read upset for an accessed SRAM cell is reduced because of the reduction in the noise source VDD × $C_{LBL,\ reduced}$. In the beginning of the READ operation the word line activation signal is significantly below its final value of VDD. By the time this final value is reached, the bit-line has been discharged significantly. This is how the effective bit-line capacitance for the accessed SRAM cell is decreased and it enhances the SRAM cell stability. The SRAM cell can be sized for improving the Iread and the WM (Fig. 4.1).

READ operation:

$$T_{\mathrm{BL}} \cong \frac{(C_{\mathrm{LBL}} \times \mathrm{VDD} + C_{\mathrm{GBL}} \times \Delta V_{\mathrm{min}})}{I_{\mathrm{read,cell}}}$$

$$E_{\mathrm{BL,read}} = C_{\mathrm{LBL}} \times \mathrm{VDD}^2 + C_{\mathrm{GBL}} \times \Delta V_{\mathrm{min}} \times \mathrm{VDD}$$

T—number of cells in the local hierarchy$C_{\mathrm{LBL}} \cong T \times C_{\mathrm{LBL,cell}}$

$$T_{BL} \cong \frac{(T \times C_{\mathrm{LBL,cell}} + C_{\mathrm{GBL}}) \times \Delta V_{\mathrm{min}}}{I_{\mathrm{read,cell}}}$$

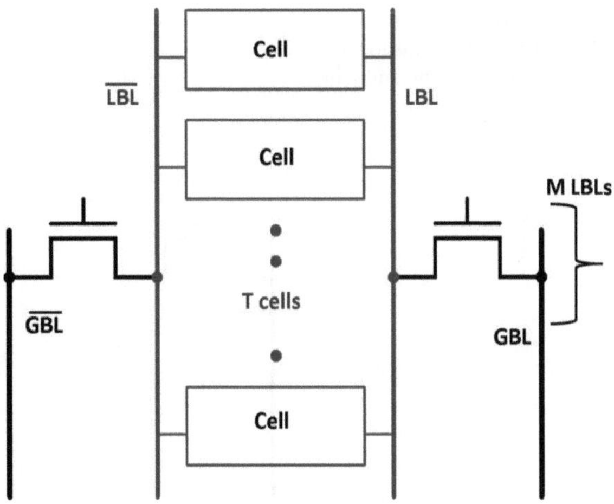

Fig. 4.1 Hierarchical divided bit-lines [Kar98]

$$E_{\text{BL,read}} = (T \times C_{\text{LBL,cell}} + C_{\text{GBL}}) \times \Delta V_{\min} \times \text{VDD}$$

WRITE operation:

$$E_{\text{BL,write}} = (T \times C_{\text{LBL,cell}} + C_{\text{GBL}}) \times \text{VDD}^2$$

During READ operation, the voltage swing of the bit-lines is limited to a smaller value (depending on the resolution of the SA employed) whereas a WRITE operation requires full voltage swing on the bit-lines. The charging and discharging of the highly capacitive bit-lines account for the major proportion of the dynamic energy consumption. This result in energy consumption of the WRITE operation to be much higher compared to the READ operation. The WRITE operation dynamic energy consumption is a critical parameter in the design of the ultra low energy SRAMs. (Yang and Kim 2005) proposed a low power WRITE operation with the use of local write receiver in local bit-line hierarchy (Fig. 4.2) The WRITE operation is executed with low swing data transfers from high capacitive global bit-lines onto much less capacitive local bit-lines where the full swing conversion is done by the local write receiver.

$$E_{\text{BL,write}} = C_{\text{LBL}} \times \text{VDD}^2 + C_{\text{GBL}} \times Vddl \times Vddl$$

$$E_{\text{BL,write}} = C_{\text{LBL}} \times \text{VDD}^2 + C_{\text{GBL}} \times \text{VDD}/4 \times \text{VDD}/4$$

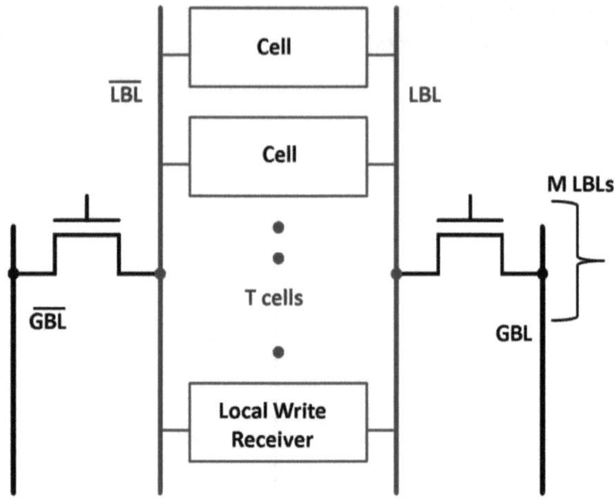

Fig. 4.2 Hierarchical divided bit-lines with local write receiver (Yang and Kim 2005)

4.3 Hierarchical Divided Bit-Lines with Local Assist Circuitry

The hierarchical bit-lines reduce the effective bit-line capacitance and the upsized low Vt local bit-line/global bit-line access transistor employed delivers more read current. The relieved small sized SRAM cell can then be optimized for improving the read SNM.

The bit-line sensing delay (T_{BL}) is

$$T_{BL} = Td_{LRBL} + Td_{GRBL} \tag{4.1}$$

Td_{LRBL} local bit-line sensing delay
Td_{GRBL} global bit-line sensing delay

$$T_{BL} = Q_{LRBL}/I_{read,cell} + Q_{GRBL}/I_{Access} \tag{4.2}$$

$I_{read,cell}$ the read current of SRAM cell and determines Td_{LRBL}
I_{Access} the current delivered by the upsized low Vt access transistor

$$Q_{LRBL} = C_{LRBL} \times V_{swing}$$

$$C_{LRBL} = C_{LRBL,cell} \times T$$

$$Q_{GRBL} = C_{GRBL} \times \Delta V_{min}$$

V_{swing} the local bit-line discharge required to trigger the local access transistor

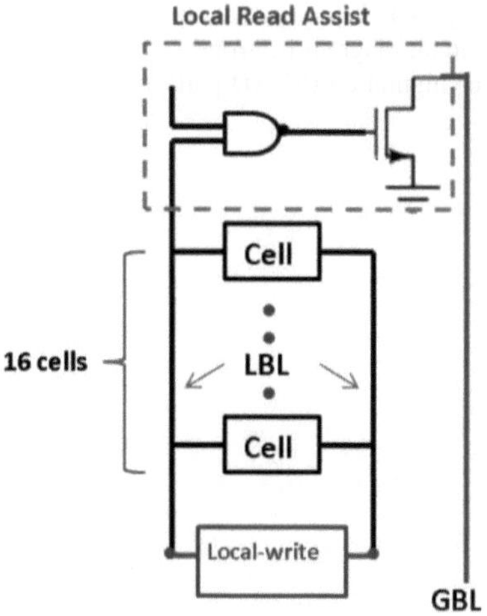

Fig. 4.3 Kawasumi et al. 2008, the array configuration. Hierarchically divided bit-lines with shared local read assist circuit between 2 local blocks. Each local block consists of 16 SRAM cells

ΔV_{min}	the minimum voltage swing required on the global bit-line required to trigger the sense amplifier
T	number of cells in local bit-line hierarchy
$C_{\mathrm{LRBL,cell}}$	the local bit-line capacitance per cell

Substituting the values in Eq. (4.2)

$$T_{\mathrm{BL}} = C_{\mathrm{LRBL,cell}} \times T \times V_{\mathrm{swing}}/I_{\mathrm{read,cell}} + C_{\mathrm{GRBL}} \times \Delta V_{\mathrm{min}}/I_{\mathrm{Access}} \quad (4.3)$$

The energy consumption per access is defined as

$$E_{\mathrm{BL,read}} = E_{\mathrm{LRBL}} + E_{\mathrm{GRBL}} \quad (4.4)$$

$V_{\mathrm{swing}} = \mathrm{VDD}$ (full swing local bit-line sensing)

$$E_{\mathrm{BL,read}} = C_{\mathrm{LRBL,cell}} \times T \times \mathrm{VDD}^2 + C_{\mathrm{GRBL}} \times \mathrm{VDD} \times \Delta V_{\mathrm{min}} \quad (4.5)$$

4.3.1 Fine Grained Bit-Line Architecture

Kawasumi et al. (2008) utilizes hierarchical bit-lines with an asymmetrical SRAM 6T cell to achieve high speed, low voltage operation. It uses asymmetrical SRAM cell with unit beta ratio with hierarchical divided bit-lines. Figure 4.3 shows local bit-line array configuration. Hierarchically divided bit-lines with 16 SRAM cells on the local bit-lines increase the SRAM cell stability. The SRAM cell can then be designed for enhanced write margin and the cell read current. The write margin is improved by increasing the drive ratio of the access transistor to the pull up

Fig. 4.4 Ishikura et al. 2008, circuit diagram of array configuration (READ part)

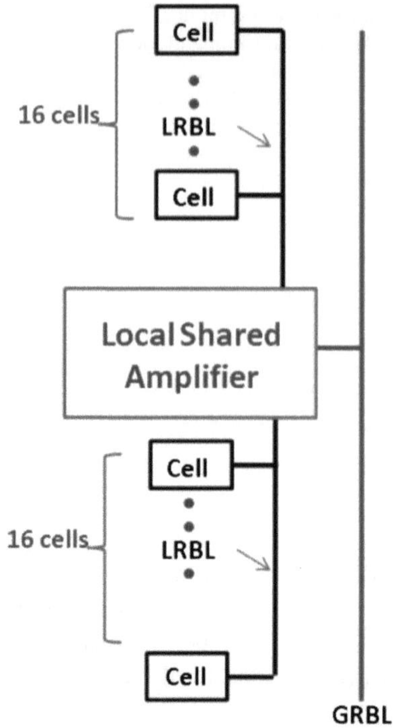

transistor. The access transistor is designed to have exactly the same size as the pull down transistor ($\beta = 1$). The design utilizes single ended sensing. The access time (local bit-line sensing) is further reduced by increasing the drive strength of the access transistor of the corresponding side of the SRAM cell connected to the local read buffer (asymmetrical SRAM cell). The asymmetry introduced in the SRAM cell does not degrade the cell stability and the WM because the variability reduction due to the transistor enlargement and dominates the influence of the symmetry distortion. A test chip of 64 kb SRAM macro in 45 nm achieves VDDmin of 0.7 V operating at 1 GHz.

4.3.2 *Divided Read Bit-Line and Read End Detecting Replica Circuit*

Chang et al. (2008) achieves high speed READ operation by replacing SRAM 6T cell with a dual port 8T cell in the local hierarchy.

The read bit-line is shared for 8 SRAM 8T cells. Ishikura et al. (2008) proposes high speed simultaneous READ/WRITE operation by replacing SRAM 6T cell with dual port 8T cell in the local hierarchy. The read bit-lines are divided into 32 local bit-lines which contains 16 memory cells. The local amplifier consists of

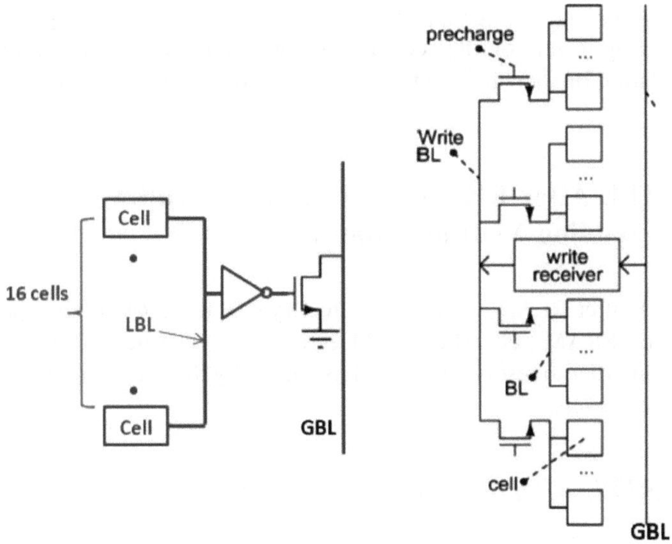

Fig. 4.5 Short buffered bit-line with shared write receivers (Cosemans et al. 2007)

domino circuit and is shared between the two local bit-line blocks. The output of local amplifier is connected to the global read bit-lines. The write bit-lines are not divided to minimize the area penalty. Figure 4.4 shows the circuit diagram of divided bit-lines with a shared local amplifier.

The leakage current generation because of the simultaneous READ and WRITE access at the same row, raising the internal nodes result in the misreading. This is remedied with the read end replica (RER) circuit. The read end replica circuit consists of the read path replica which is used to adjust the timing of the global amplifiers for the normal worst case reading scenario and the misreading. For the normal case bit-lines are driven by the complete "ON" state transistors. Therefore, normal reading occurs faster than misreading (due to partial raised internal nodes). The RER circuit immediately closes timing event after the normal reading and the misreading is prevented. The test chip (Ishikura et al. 2008) of 64 kb SRAM macro in 45 nm LP achieves VDDmin of 0.75 V. For the worst process condition access time is 1.9 ns at VDD of 1.0 V.

4.3.3 Short Buffered Local Bit-Lines with Low Swing GBLs

Cosemans et al. (2007) proposes the low swing global bit-lines during the READ and WRITE operation. The local write receivers are shared between several local bit-line blocks, thereby reducing the area penalty. The local read buffer is shared between the 2 local bit-lines block (16 SRAM 6T cells per local bit-line block). The global bit-lines are pre-charged to a lower value of 200 mV. Thus the reduced value of pre-charge voltage enables well-controlled voltage swing on the GBL without having to rely on the accurate timing requirement. Figure 4.5 shows the implementation of short buffered local bit-lines with low swing global bit-lines.

4.4 WRITE After READ Based Assist Circuitry for Enabling VDDmin Operation.

The WRITE after READ based assist circuitry enables DRAM-type operation for the accessed SRAM cell. The SRAM cell data is rewritten after every READ operation, in order to resolve the data flips caused by the cell stability issues at low VDD voltage levels.

4.4.1 WRITE After READ Based Assist Circuitry

The acceleration of the bit-line discharge process, limits the amount of charge injected from the VDD pre-charged bit-line to the low node of the accessed SRAM cell. This enhances the SRAM cell stability (Pilo et al. 2007). The data refreshing (writing back of data) after every read operation with help of a sense amplifier virtually eliminates the cell stability issue. Pilo et al. 2007 proposes an integration of a sense amplifier on each column of the sub array of size 64 bits by 192 bits. The sense amplifier enables full bit-line amplification, accelerating the discharge of the low node of the cell, and also results in the data recovery of the original data of the SRAM cell. The full bit-line amplification for the sub array (64×192 bits) results in very high energy consumption. For example, full amplification on the sub array of 64×192 bits increases the energy consumption by 12 times compared to the required 100 mV bit-line discharge (normal READ operation) at VDD = 1.2 V. Pilo et al. 2007 solves this issue by masking the sense amplifier activation for the sub array columns that have sufficient functionality margins. A test chip macro of 32 Mb in 65 nm LP achieves VDDmin of 0.8 V.

4.4.2 Short Buffered Bit-line

The energy consumption associated with WRITE after READ operation can also be reduced by reducing the full swing switching bit-line capacitance (Cosemans et al. 2009). Cosemans et al. 2009 proposes short buffered bit-lines with 16 cells per local bit-line. The capacitance switched for write operation after read operation is reduced which results in the reduced energy consumption. Each local bit-line is split in 4 sub-local bit-lines and local write receiver is shared for these 4 sub-local bit-lines. A test chip macro of 128 Kb in 90 nm achieves VDDmin of 1.0 V operating at 240 MHz.

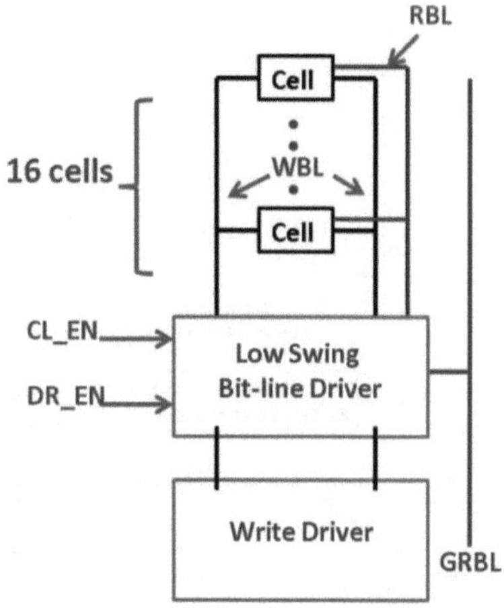

Fig. 4.6 Array configuration of low energy disturb mitigation technique (Yoshimoto et.al. 2011)

4.4.3 Low-Energy Disturb Mitigation (Half Select Issues) Scheme

Yoshimoto et al. 2011 proposes a write back scheme to overcome the half select problem for 8T SRAM cell based arrays. The proposed scheme consists of a floating bit-line technique and a low swing bit-line driver. Figure 4.6 shows array configuration of the proposed scheme. For an activated column CL_EN (column line enable) signal is enabled, DR_EN (driver disable signal) is "high" and the write driver drives the write bit-line. For the columns having half-selected cells CL_EN is low and the DR_EN is pulled down in the selected row. The low swing bit-line driver gets activated and pulls up or pulls down each write bit-line. With the result the write bit-lines are floating both for the activated mode and the standby mode. A test chip of 512 Kb 8T SRAM macro achieves a VDDmin of 0.5 V.

WRITE after READ based circuit techniques resolve the cell stability issues with the result that the SRAM cell can be sized favoring the write margin improvement and the enhanced read current. WRITE after READ based circuit techniques can be an alternative enabling high speed READ access. The extra energy wasted in performing full swing WRITE operation after every READ operation makes it less optimum choice for enabling low energy READ access.

4.5 Low Swing Bit-Line Hierarchy: Enhanced SRAM Cell Stability

The hierarchically divided local bit-lines with local assist circuitry as discussed earlier has a major drawback in achieving ultra low energy read access. First, the low Vt read buffers used increase the leakage power. Second, the read buffers used are operating with the full swing pre-charged local bit-lines (LBLs), resulting in an

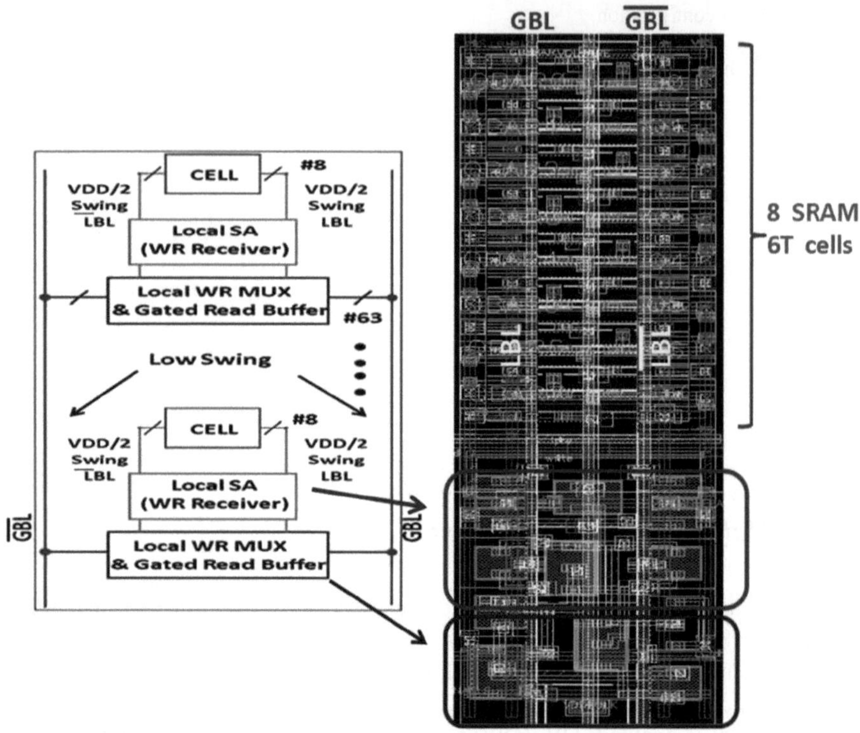

Fig. 4.7 Pseudo 8T sensing enabled local assist Circuitry: VDD/2 pre-charged local bit-lines (Sharma et.al. 2010)

increased dynamic energy consumption. This problem is addressed by a novel local assist circuit which enables low swing local bit-lines and solves the leakage issue by using gated read buffers. Pseudo 8T sensing enabled local assist circuitry (Sharma et.al. 2010) remedies the above-mentioned issues and offers higher cell stability. Hierarchical buffered segmented bit-lines (Sharma et al. 2011) achieve the high performance, low energy, and higher cell stability without resorting to energy expensive WRITE after READ mechanism.

4.5.1 Pseudo 8T Sensing Enabled Local Assist Circuitry

The design optimization problem of 6T SRAM cell, improvement in read SNM without the degradation of access time is solved with this architecture (Sharma et al. 2010). The local bit slice architecture is shown in Fig. 4.7 features:

1. A local sense amplifier which also acts as a local write receiver during a WRITE operation.

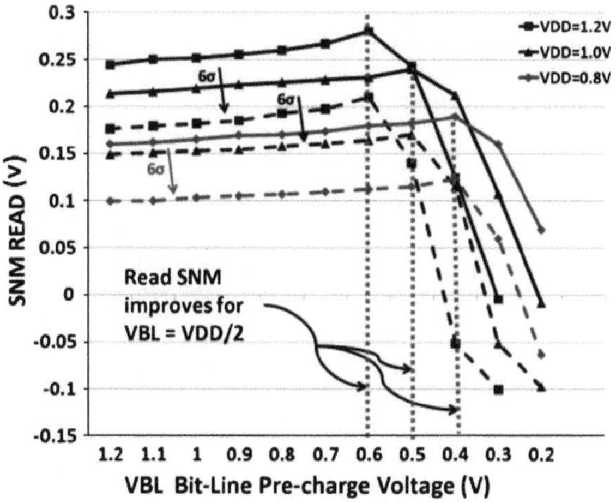

Fig. 4.8 Read SNM (cell stability) versus bit-line pre-charge voltage for the HVT minimum sized transistors based SRAM cell (DC simulations)

2. VDD/2 pre-charged local bit-lines resulting in an increased read SNM and also results in charge recycling, thereby reducing dynamic energy consumption.

3. Gated read buffers deliver the required read current and also mitigates the local bit-line leakage.

There are 8 number of SRAM cells in the local hierarchy. The high Vt transistors based SRAM cell reduces leakage and enhances SRAM stability. The local sense amplifier relieves the accessed SRAM cell from creating enough voltage swing required to trigger the read buffer. The required access current is delivered by an upsized gated read buffer. This buffer is enabled, only for a limited period during the READ operation thereby reducing the local bit-line leakage. The energy reduction perspective of this architecture will be discussed in Chap. 5.

Impact of VDD/2 pre-charged local bit-lines on the SRAM cell stability

The VDD/2 pre-charge value for the short local bit-lines further improves cell stability, Fig. 4.8. The cell side that stores "1" gets discharged for the pre-charge voltages less than VDD_{WL}—Vt, resulting in the read upset failures. But for the HVT transistors based SRAM cell, VDD/2 pre-charge value for the bit-lines is greater than VDD_{WL}—Vt. Therefore, the cell stability degradation for the reduced bit-line pre-charge voltage is valid for the bit-line voltages less than VDD/2.

4.5.2 Hierarchical Buffered Segmented Bit-Lines

Hierarchical buffered segmented (HBS) bit-line is a low swing based local bit-line technique enabling high performance SRAM (Sharma et al. 2011). HBS based local bit-line architecture involves segmentation of the local bit-lines. The segment buffers are inserted in each local hierarchy (Fig. 4.9). The local bit-line connecting

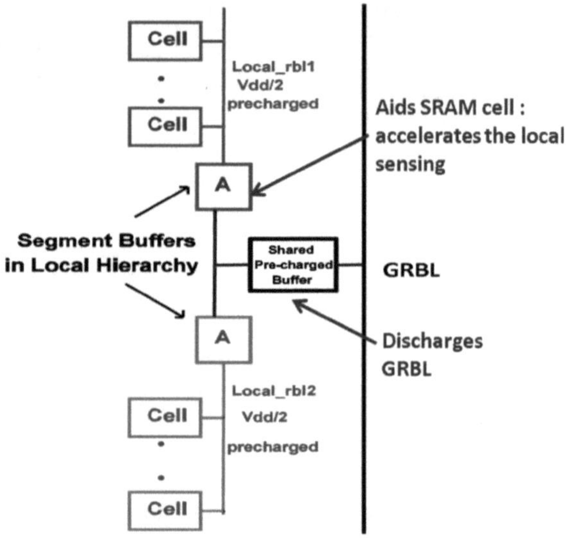

Fig. 4.9 Hierarchical Buffered Segmented bit-lines based SRAM (Vddl = vdd/2) (Sharma et al. 2011)

to the SRAM cells is pre-charged to a lower value, say VDDL (e.g. VDD/2) thereby reducing the active power consumption associated with pre-charging and discharging and increasing the SRAM cell stability (Fig. 4.9). The lower value of the local bit-line pre-charge value is selected such that the overdrive of the NMOS transistor (segment buffer) remains negative for the accessed SRAM cell storing "Low" value (corresponding bit-lines remain pre-charged for the "Low" values stored accessed SRAM cells). The segment buffers are enabled after the local bit-line is discharged low enough. This further increases the SRAM cell stability as the activated SRAM cell on the local bit-lines during the initial phase of READ operation is isolated from the extra parasitic capacitance of the local assist circuitry (read buffer). The parasitic capacitance of the local assist circuitry cannot be neglected for the short local bit-lines (8–16 cells per local bit-line). This parasitic capacitance is an extra load on the local bit-lines which prevents faster swing of the bit-line, thereby impacting the cycle time and also the cell stability.

The read buffer consists of global access transistor (Fig. 4.10a) driven by an inverter, held at logic low when not used by pre-charging the input to VDD. In order to limit the number of pre-charged read buffers, each one is shared with 2 local blocks. The pre-charged read buffer is driven by the segment buffer of an activated local block. The usage of segment buffers improves the performance compared to the conventional hierarchical divided full swing architectures. The Iread created by an accessed minimum sized SRAM cell is amplified by the segment buffer (Fig. 4.10a). The size of the NMOS transistor used is 2x the size of the transistors used in SRAM cells. The area penalty associated with up sizing of this NMOS transistor is much less as it is shared with 8–16 SRAM cells used in the local hierarchy. The toggling of the read buffer depends on the rate of discharge of node (In). In the case of hierarchical divided full swing bit-lines (Kar98, Cosemans et al. 2007; Chang et al. 2008;

Ishikura et al. 2008] node (In) is discharged by the Iread, cell. Whereas, with the use of segment buffers the node (In) is discharged by A x Iread, cell which results in faster toggling of the read buffer (Fig. 4.12b). During an idle state the local bit-lines are pre-charged at VDD/2 and the input node of pre-charged read buffer is held high (VDD) and the global access transistor are kept OFF.

$$\begin{aligned} T_{\text{BL,delay}}(HBS) &= T_{\text{local\_rbl}} + T_{\text{GRBL}} \\ &= \frac{Q_{\text{local\_rbl}}}{I_{\text{local\_rbl}}} + \frac{Q_{GRBL}}{\text{I}_{\text{Access}}} \\ &= \frac{C_{local\_rbl} \times \text{V}_{swing}}{(A)I_{\text{read,cell}}} + \frac{C_{\text{GRBL}} \times \Delta\text{v}_{\text{min}}}{I_{\text{Access}}} \end{aligned}$$

$$E_{\text{BL}}(HBS) = E_{\text{local\_rbl}} + E_{\text{GRBL}}$$

$$E_{\text{local\_rbl}} = T \times C_{\text{local\_rbl\_cell}} \times (V_{\text{local\_rbl}} \times V_{\text{swing}})$$

$$E_{\text{GRBL}} = C_{\text{GRBL}} \times (V_{\text{GRBL}} \times \Delta V_{\text{min}})$$

V_{swing} minimum voltage swing required on the local bit-line for toggling of the read buffer.
$I_{\text{read,cell}}$ SRAM cell read current.
$C_{\text{local\_rbl}}$ capacitance of local read bit-line
C_{GRBL} capacitance of global read bit-line.
I_{Access} drain current of global access transistors.
ΔV_{min} minimum voltage difference to be resolved by the sense amplifiers
T number of cells in the local hierarchy.

4.6 High Bit Density Based Bit-Line Hierarchy

The use of divided bit-lines with local assist circuitry improves the cell stability and reduces access time. But the use of local assist circuitry with hierarchical bit-lines requires a lot of dummy cells and extra metal resources for routing additional global bit-lines.

This severely degrades SRAM cell density. This section discusses an alternative local bit-line hierarchy which enables a lower supply voltage combined with the higher cell density.

4.6.1 Cascaded Bit-Line with Self-Write-Back Sense Amplifier

The subarrays are connected with each other by the capacitance separators, activated by CS_En_i (Fig. 4.11, kushida et al. 2009). The data from and to the memory cells in the subarrays are transferred through the cascaded bit-lines.

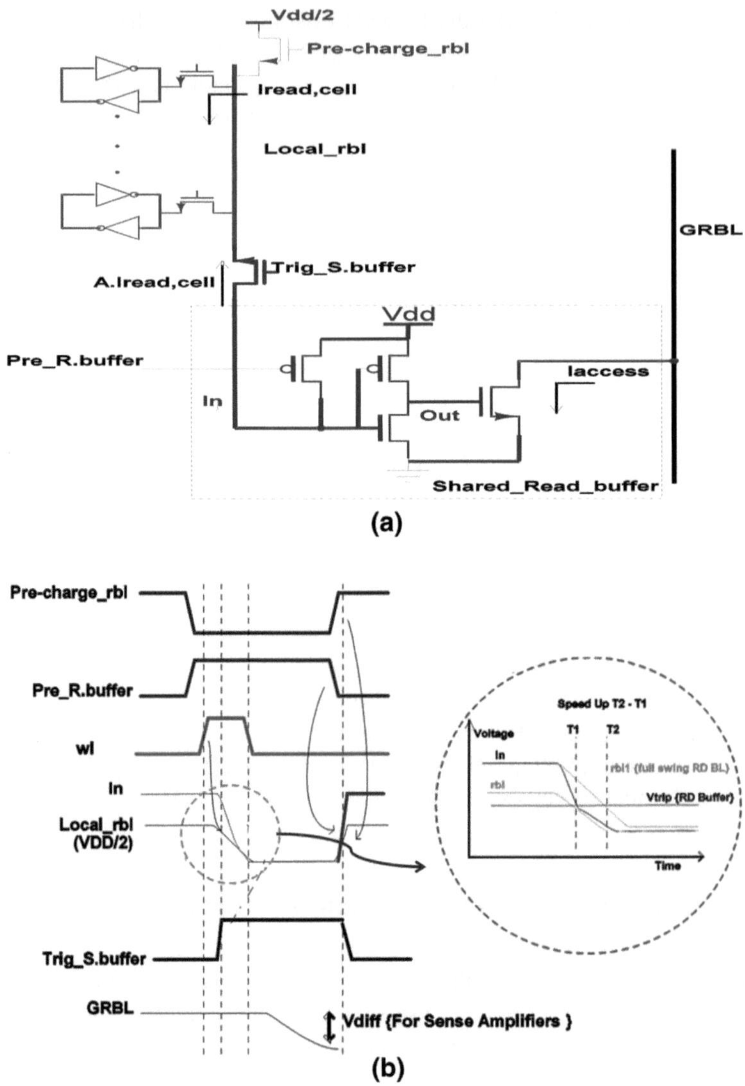

Fig. 4.10 **a** Circuit diagram of HBS based 6T SRAM. **b** READ Access Operation for HBS based 6T SRAM cell {reading "L"}

The capacitance separator isolates parasitic capacitance of the local assist circuitry and protects against the disturbing effect of parasitic capacitance in the accessed SRAM cell. This enables the rapid discharging of a bit-line. The capacitance separator is activated, after when the bit-line has discharged the bit-line low enough. This transfers the bit-line information to the input node of the sense amplifier. The sense amplifier is then triggered by the discharge information and the latching process gets initiated and writes back the cell data. This sequential

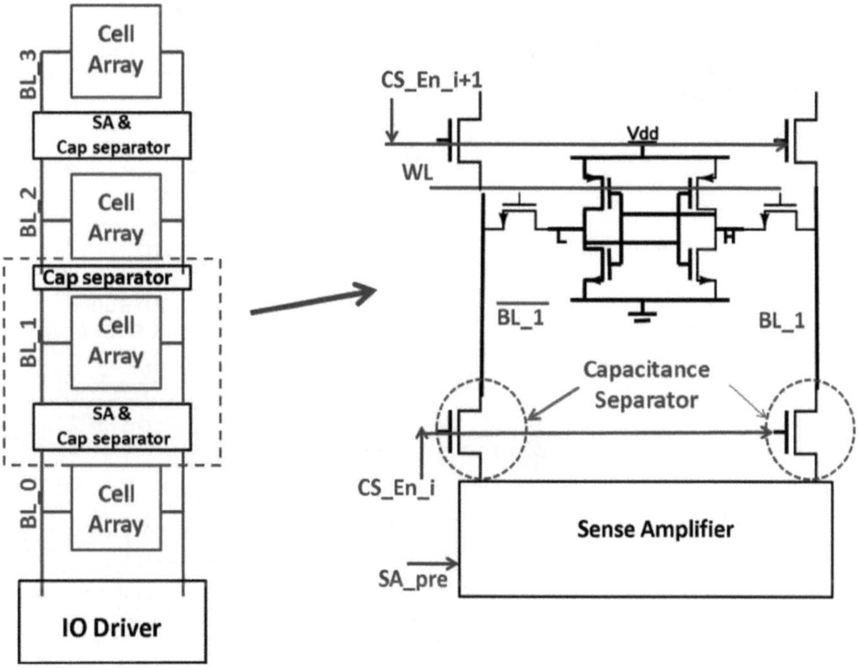

Fig. 4.11 Concept and architecture of cascaded bit-lines

operation is continued till the last sense amplifier of the activated column which then transfers the data to input/output circuitry. The proposed SRAM macro in 65 nm achieves READ/WRITE functionality at 0.7 V with very high cell density of 0.56 Mb/mm^2.

The energy consumption of cascaded bit-lines is very high as the read cycle with the proposed design requires full swing read bit-lines and second the complex timing control logic required to sequential triggering of the capacitance separator of the accessed columns, increases the energy consumption. Third the write-back for SRAM cell stability (similar to the refresh in DRAM) as discussed earlier is not an optimum choice from the energy consumption perspective.

4.6.2 SRAM Cell Type Local Assist Circuitry

An 8T SRAM cell type structure of the local assist circuitry (NS-LWR, WR MUX, and local read buffer, Fig. 4.12) results easy implementation of SRAM cell dense layout rules (litho friendly) implementation (Sharma et al. 2012). The physical regularity of SRAM layout enables the use of litho optimized specialized DRC rules. The advantage of ultra regular layout of SRAM matrix in achieving area

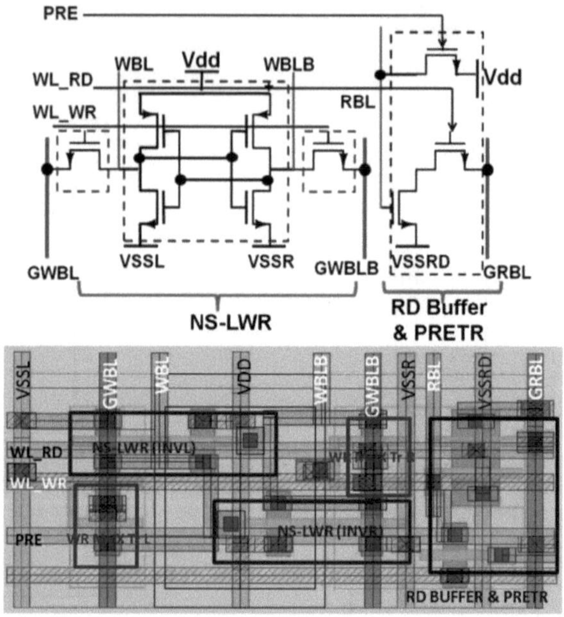

Fig. 4.12 8T SRAM cell type layout of the local assist circuitry: ~2x of actual 8T SRAM cell (Sharma et al. 2012)

reduction is quite obvious. But achieving the same benefit from the logic circuit is difficult because of the irregularity in the logic circuit layout. As a result the conventional logic circuit based local assist techniques complicate the litho optimization of the memory matrix.

Also the 2 cross coupled inverters reduce the timing complexity associated with the strobe signal generation of the local write receiver. The WL_WR activation signal for the WR pass transistors transfers the low swing information onto the local bit-lines and also serves the purpose of triggering the regenerative action of the 2 cross coupled inverters. The pulsed WL_WR signal isolates the nodes of the cross coupled pair from the highly capacitive bit-lines. This architecture also enables the differential VSS biasing technique, which allows the independent tuning of the VSS connection of the cross coupled inverters of SRAM cells and the NS-LWR.

The SRAM cells of the local hierarchy and NS-LWR has connections to left and right vertical VSS rails VSSL and VSSR. The data dependent bias application on VSSL and VSSR for the offset cancellation of the local write receiver also improves write-ability of the accessed SRAM cells as discussed in the Chap. 2.

The sources of 8T SRAM cells read buffers and local read buffer (upsized LVT transistor) are both connected to VSSRD, which is kept floating for all non-accessed matrix columns. The floating VSSRD and low swing pre-charge voltage for GRBL reduces the leakage current by 40x for the worst case (fast NMOS and fast PMOS process corner). The matrix column for an accessed SRAM cell is activated by connecting its VSSRD port to the GND. The asserted 8T

SRAM cell discharges the local read bit-line depending on the stored data information. Then the local read buffer is activated by WL_RD signal. Local read buffer transfers the information from the local read bit-line to GRBL to be sensed by the global sense amplifiers.

The local assist circuitry as proposed in this work consisting of NS-LWR, WR pass transistors, and local read buffers are easy to map onto regular design fabric, similar to SRAM cells. The components of the local assist circuitry consisting of 2 cross coupled inverters of the local write receiver, 2 NMOS pass transistor of the WR MUX, and the 2 stack NMOS transistor of the local read buffer resembles an 8T SRAM cell. The additional NMOS pre-charge transistor for the local read bit-line is implemented in the local read buffer region (Fig. 4.12). In other words this local assist circuitry facilitates shape–level regularity requirement to take advantage from the litho optimization. Otherwise enforcing shape–level regularity for litho optimization is a difficult task with the existing conventional local assist techniques. Therefore the 8T SRAM cell type implementation of the proposed local assist circuitry offers enhanced flexibility for embedding the logic circuit into the memory matrix at a reduced area cost.

4.7 Comparative Analysis

4.7.1 Performance

The SRAM cell current is getting considerably reduced in advanced sub nanometric technology nodes due to the transistor scaling and the Vt random variations. With the result the memory access times is severely impacted. Figures 4.13 and 4.14 show the bit-line delay for the memory of size 512 $\times$ 512 cells in 65nm and 40 nm for the nominal process corner. For the fair comparison only minimum sized Wmin standard Vt based SRAM 6T cells are used for different local architectures. The BL delay is defined as the time from the WL activation to the voltage drop of 150 mV on the global bit-lines. The number of cells taken in each local hierarchy is 16, except for Chang et al. (2008). The number of cells proposed for the local architecture is 8 in Chang et al. (2008). The transistor size for the read buffers taken is 2x Wmin. For HBS bit-lines (Sharma et al. 2011), the size of segment buffers and pre–charged read buffers is also 2x Wmin. The local assist circuit based hierarchical divided bit-lines (Ishikura et al. 2008; Chang et al. 2008; Cosemans et al. 2007) offers higher performance. The reduced effective bit-line capacitance (8–16 SRAM cells on a local bit-line) and a global bit-line discharge by an enhanced access current delivered by an upsized local read buffer results in high performance. The reduced pre-charged global bit-lines as proposed by Cosemans et al. 2007 reduces the dynamic bit-line switching energy and offers a higher degree of variability resilience by limiting the amount of maximum charge available for discharge. But the reduced pre-charged global bit-line results in a

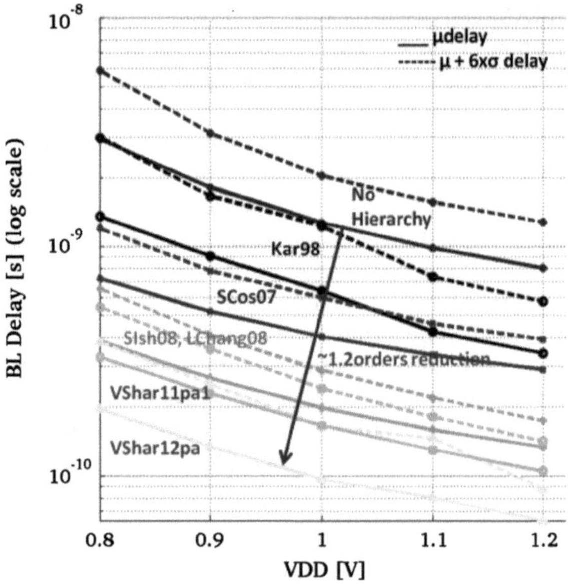

Fig. 4.13 BL Delay [150 mV bit-line discharge from word line assertion] versus VDD for different local architectures. Column height is 512 cells, 65 nm LP, nominal process corner & 25 °C

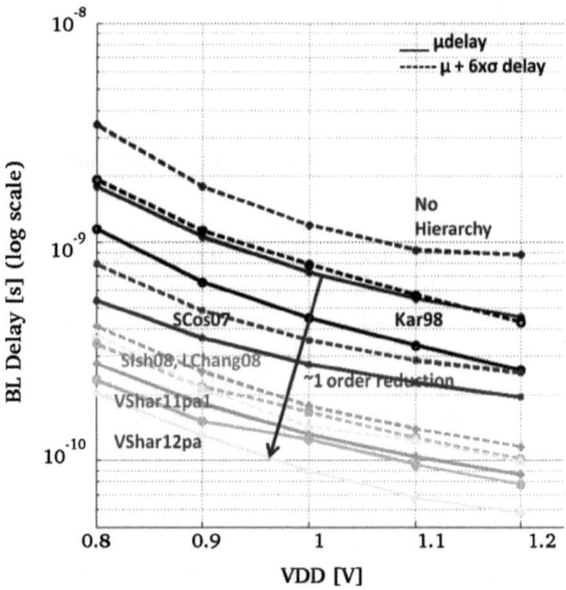

Fig. 4.14 BL Delay [150 mV bit-line discharge from word line assertion] versus VDD for different local architectures. Column height is 512 cells, 40 nm LP, nominal process corner & 25 °C

performance loss compared to the VDD pre-charged global bit-line. Therefore, the performance of Cosemans et al. 2007 is less compared to Ishikura et al. 2008 and Chang et al. 2008.

The HBS architecture offers higher performance for all the VDD values from 0.8 to 1.2 V. The HBS architecture (Sharma et al. 2011) based SRAM cell design

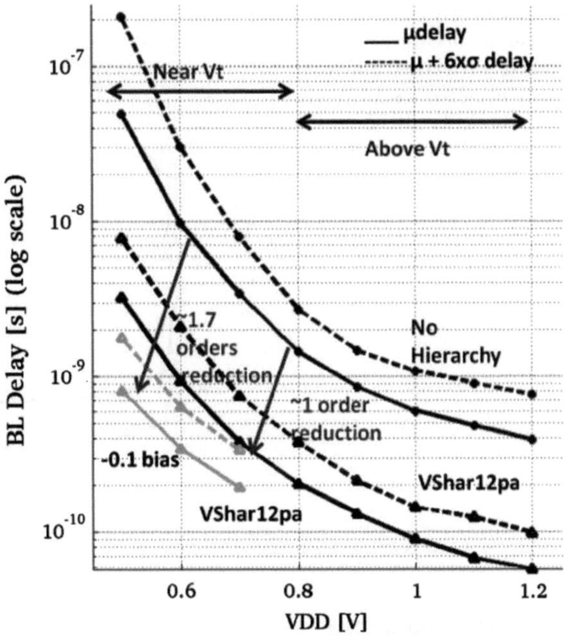

Fig. 4.15 BL Delay [150 mV bit-line discharge from word line assertion] versus VDD for litho optimized local assist circuitry (Sharma et al. 2012). Column height is 512 cells, 40 nm LP, nominal process corner and 25 °C

is (8–9)x faster compared to the conventional SRAM design (no hierarchy) in 65 nm (Fig. 4.13) and (6–8)x faster in 40 nm for VDD = 1.2 V. The HBS is (9–10)x faster for 65 and 40 nm at VDD = 0.8 V. The performance of HBS architecture is (1.3–1.4)x better compared to the other high performance architectures [Ishikura et al. 2008 and Chang et al. 2008]. This higher performance of HBS architecture is because of the segment buffers in a local hierarchy. The insertion of segment buffer isolates the parasitic capacitance and further reduces the effective bit-line capacitance for an accessed SRAM cell. And also the segment buffer aids the accessed SRAM cell in discharging the local bit-line. This also helps in improving the cell stability. The low Vt SRAM cell with HBS offers the same cell stability compared to the standard Vt SRAM cell on the short local bit-lines. However, for this analysis only standard Vt SRAM cells are used but in actual, the performance gain with HBS is even more better.

The SRAM cell type local assist circuitry (Sharma et al. 2012) reduces the area overhead and reduces the bit-line parasitic capacitance which reduces the bit-line delay. This map into 1.2 orders of reduction in bit-line delay for 65 nm (Fig. 4.13) and 1 order of reduction for 40 nm (Fig. 4.14). The bit-line delay can further be reduced to 1.7 orders of magnitude (Fig. 4.15) for the scaled voltage levels for 40 nm with the application of −0.1 V biasing on VSSRD (the source of local read buffer). Therefore, the litho optimized local assist circuitry results in an overwhelming improvement in the access speed for the scaled voltage levels.

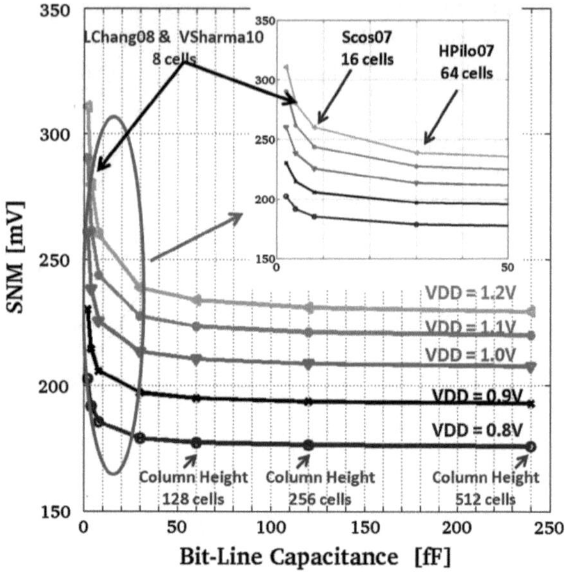

Fig. 4.16 Dynamic SNM versus bit-line capacitance for different VDD and different local architectures. Column height is 512 cells, 65 nm LP, nominal process corner and 25 °C

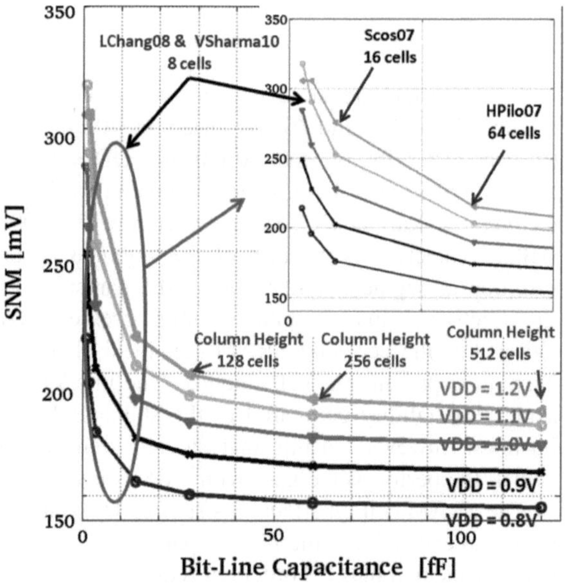

Fig. 4.17 Dynamic SNM versus bit-line capacitance for different VDD and different local architectures. Column height is 512 cells, 40 nm LP, nominal process corner and 25 °C

4.7.2 Stability Analysis

The effective bit-line capacitance reduction with hierarchical divided bit-line architecture results in enhanced SRAM cell stability. Figures 4.16 and 4.17 show dynamic SNM read versus bit-line capacitance for different bit-line capacitances in

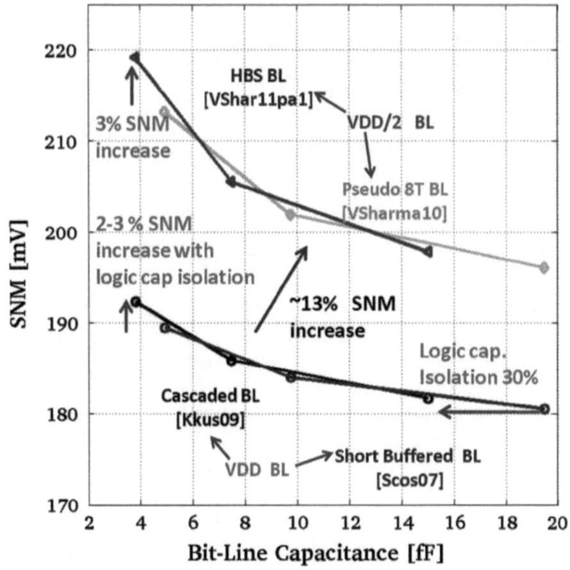

Fig. 4.18 Dynamic SNM versus bit-line capacitance of enhanced dynamic SNM based architectures. Column height is 512 cells, 65 nm LP, nominal process corner and 25 °C

65 and 40 nm for the nominal process corner. The minimum sized Wmin standard Vt based SRAM 6T cells are used for different local architectures for dynamic SNM analysis. The reduction in the number of SRAM cells on the matrix column (short column height) reduces the effective bit-line capacitance. This reduces the magnitude of the bit-line charge (noise source Q = VpreBL × Cbl) and increases the cell stability.

The improvement in dynamic SNM is the maximum for the architectures supporting 8 SRAM cells on the local bit-line hierarchy, for example, Chang et al. 2008 and Sharma et al. 2010. The dynamic SNM can be further improved by further reducing the noise source (bit-line charge) by reducing the local bit-line pre-charge voltage. The VDD/2 pre-charged short local bit-lines (Sharma et.al. 2010) results in 13 % and 20 % improvement in SNM read for 65nm and 40 nm (Figs. 4.18 and 4.19) resulting in an increased read SNM and reduction in the dynamic energy.

The parasitic capacitance of logic assist circuitry used with hierarchically divided bit-lines increases the effective bit-line capacitance. The increment is not negligible for a short bit-line, such as 8–16 cells per bit-line. This additional increase in the capacitance prevents faster bit-line discharge by an accessed SRAM cell and degrades the cell stability. This problem is remedied by isolating the accessed SRAM cell from the parasitic capacitance of the local assist circuitry as proposed by Kushida 2009 and Sharma et al. 2011.

The capacitance separator (isolates the parasitic capacitance) (Kushida 2009) immunizes the disturbing effect of parasitic capacitance in the accessed cell. In the beginning of a read cycle, the local bit-line swings rapidly because of the isolation of the local assist circuitry capacitance (approximately 30 %) compared to the

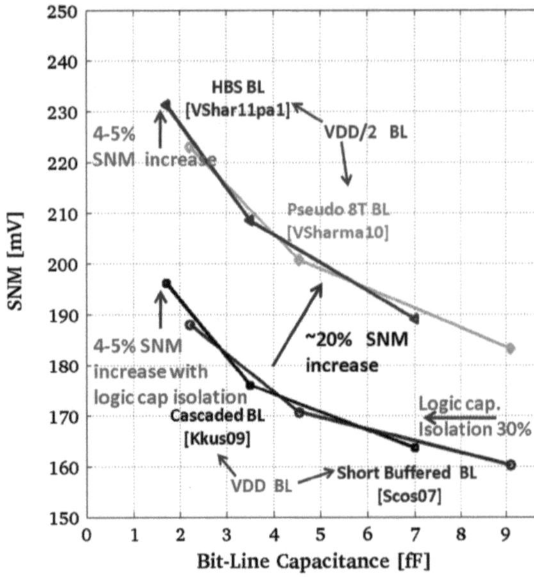

Fig. 4.19 Dynamic SNM versus bit-line capacitance of enhanced dynamic SNM based architectures. Column height is 512 cells, 40 nm LP, nominal process corner and 25 °C

scenario when there this no parasitic capacitance isolation. This is because the capacitance separator shields the parasitic capacitance of the local assist circuit until bit-line voltage is lowered and the capacitance separator is opened. The 30 % parasitic capacitance isolation for the short bit-lines (8–16) SRAM cells result in 3–5 % improvement in the dynamic SNM read (Figs. 4.18 and 4.19). Kushida 2009 also proposes WRITE after READ mechanism to further improve the SNM read at the expense of increased energy consumption.

Sharma et al. 2011) proposes an energy efficient enhanced SNM read operation. The segment buffers inserted in the local bit-line architecture isolate the parasitic capacitance of the local assist circuitry. The segment buffers are enabled only after the local bit-line has been discharged to a predefined limit and the word line signal has been disabled. In addition to this reduced pre-charged local bit-lines (VDDL = VDD/2) also reduces the magnitude of the noise source thereby further resulting in an enhanced SNM read. The dynamic SNM is improved by approximately 23 % compared to the architectures (Cosemans et al. 2007; Ishikura et al. 2008 and Chang et al. 2008) with VDD pre-charged local bit-lines and no parasitic capacitance isolation (Fig. 4.19).

4.7.3 Energy Consumption

Energy consumption is a vital parameter in analyzing the effectiveness of an employed local assist technique. In this chapter energy consumption of only high performance SRAMs is covered. Energy efficient medium performance architectures are covered in Chap. 5. Energy efficiency of hierarchical divided bit-line architecture (Kar98; Ishikura et al. 2008 and Chang et al. 2008) can be attributed to better variability resilience and reduced effective bit-line capacitance compared to the conventional SRAM 6T cell architecture (no hierarchy). The conventional SRAM designs (highly capacitive non hierarchical bit-lines with all the SRAM cells connected) results in excessive bit-line discharge due to the process variations, thereby increasing the energy consumption. Whereas the impact of increased process variations causing excessive discharge is limited only to the local bit-lines and full swing voltage levels are used only for the local bit-lines connected to 8 or 16 SRAM cells. The access transistor of local bit-line architecture delivers more access current compared to the NMOS access transistor of SRAM 6T cell. With the result the excessive bit-line discharge by the fast transistors (positive Vt shifts) in the time dictated by the slow transistors (negative Vt shifts) is reduced. This is how the reduced impact of process variations on upsized low Vt read buffer prevents the excessive global bit-line discharge and reduces the energy consumption.

The reduction of energy consumption with litho optimized architecture (Sharma et al. 2012) is maximum, compared to the conventional hierarchical divided bit-line architecture (Chang et al. 2008 and Ishikura et al. 2008). The litho friendly SRAM cell type layout of the local assist circuitry enables compact layout, thereby reducing bit-line wire capacitances. This directly maps into reduction in energy consumption. (Sharma et al. 2012) also proposes to replace VDD pre-charged global bit-lines with reduced pre-charge voltage for further optimizing the energy consumption at the expense of decreased performance. For the same medium low performance targets, this litho friendly SRAM cell type layout reduces the energy consumption compared to the reduced global bit-line pre-charged short buffered bit-line architectures (Cosemans et al. 2007).

HBS bit-lines (Sharma et al. 2011) based SRAM design is the most energy efficient high performance architecture. The energy consumption reduction is ~22 % compared to Chang et al. 2008 and Ishikura et al. 2008 based architectures for 8 number of SRAM cells on the local bit-lines (Figs. 4.20 and 4.21). The HBS bit-line architecture is not only faster (Figs. 4.13and 4.14) but also consumes less energy compared to the Chang et al. 2008 and Ishikura et al. 2008 based SRAM designs. This energy reduction with HBS architecture is due to the VDD/2 pre-charged local bit-lines. In the conventional hierarchical divided bit-lines with local assist circuitry (Cosmans et al. 2007; Chang et al. 2008 and Ishikura et al. 2008) the voltage drop on the local bit-lines caused by the accessed SRAM cell triggers the read buffer. It is not possible to trigger the read buffer (requires full swing signals as input) with the voltage drop created on VDD/2 pre-charged bit-lines. This is remedied with HBS bit-lines architecture because the segment buffer used in the local hierarchy drives the pre-charged read buffer.

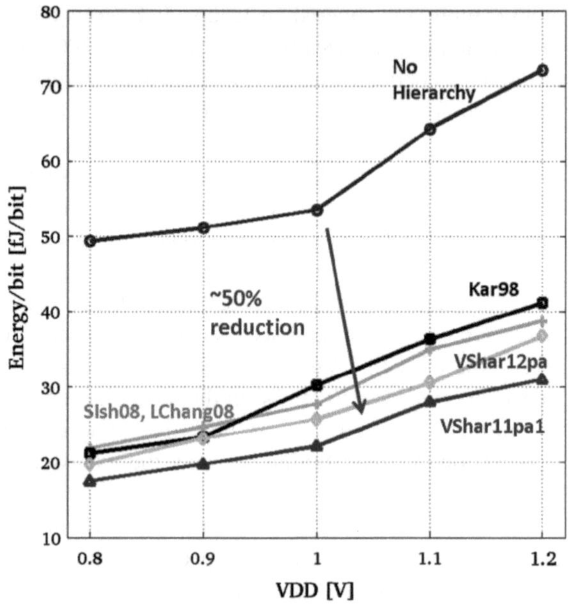

Fig. 4.20 Energy/bit versus VDD. Energy/bit is calculated for the scenario when the 99.999 % of minimum sized standard Vt based SRAM cell has discharged bit-line by the required value of 150 mV. Column height is 512 cells, 65 nm LP, nominal process corner and 25 °C

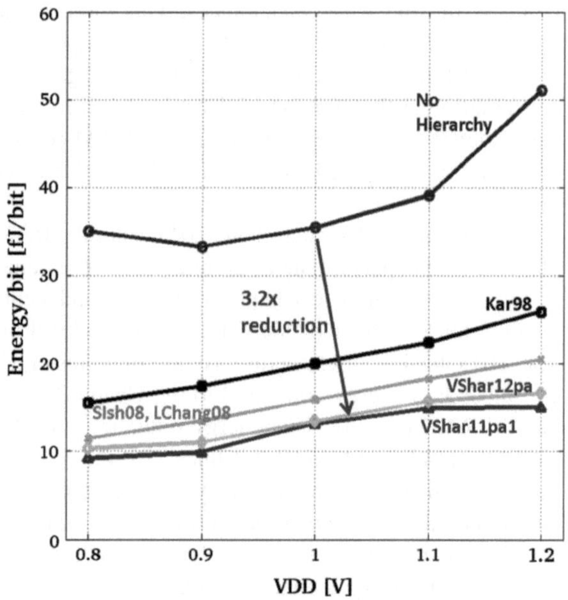

Fig. 4.21 Energy/bit versus VDD. Energy/bit is calculated for the scenario when the 99.999 % of minimum sized standard Vt based SRAM cell has discharged bit-line by the required value of 150 mV. Column height is 512 cells, 40 nm LP, nominal process corner and 25 °C

4.7.4 Area Overhead

The hierarchical divided bit-lines with local assist circuitry promises high performance, stability, and reduce the energy consumption at the cost of increased area overhead. The increased area overhead with the use of local assist circuitry

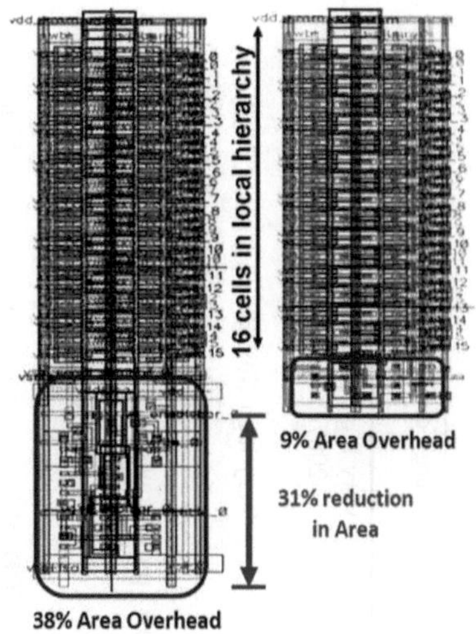

Fig. 4.22 Area overhead comparison for traditional hierarchical divided bit-line architectures versus litho optimized bit-line architecture (Sharma et al. 2012) in 65 nm LP technology

ranges from 20 % to 40 %, depending on the logic circuits used and the array configuration. This increased area overhead can be partially controlled by sharing the local assist circuitry between 2 local bit-line hierarchies as proposed by Cosmans et al. 2007, Chang et al. 2008, Ishikura et al. 2008, and Sharma et al. 2011. The use of logic circuits as local assist circuitry requires insertion of the dummy cells in the memory matrix which increases the complexity of the memory matrix optimization. There is also an issue of reduced cell density with the hierarchical divided bit-lines. The use of an additional global bit-line requires an extra metal resource which decreases the cell density. This problem is remedied by employing Cascaded BL architecture as proposed by Kushida et al. 2009. With the capacitance separator (Cascaded BL) the effective bit-line capacitance for the accessed SRAM cell is reduced. There is no requirement for the global bit-lines, so there is no need for an extra metal resource as well. However, the increased area issue is only partly addressed by just increasing the cell density by avoiding a use of an extra metal resource.

The area overhead is considerably lower for the litho optimized local assist circuitry (Sharma et al. 2012). Figure 4.22 shows the best effort layout of the litho optimized local assist circuitry versus hierarchical divided bit-lines with local assist circuitry (e.g. Cosmans et al. 2009 and Sharma et al. 2010). This work utilizes logic DRC rules based SRAM cells due to the non–availability of litho optimized parametric SRAM cells for academic purposes. The area overhead of proposed solution is only 9 % compared to 38 % with the existing hierarchical divided bit-lines with local assist circuitry (Cosmans et al. 2009 and Sharma et al. 2010). First, the non strobed local write receiver reduces the transistor count compared to the conventional strobed local write receiver. Second, DVSS bias applied for the offset mitigation further relaxes the transistor sizing requirement compared to the conventional LWR. Third, the SRAM cell type structure of non strobed LWR and associated WR MUX enables compact pitch matched layout.

Table 4.1 Comparison table for different local assist circuitry

Architecture	Performance	Cell stability (SNM read)	Energy consumption	Area overhead
HBL + local assist (Cosemans et al. 2007; Ishikura et al. 2008 and Chang et al. 2008]	High (Chang et al. 2008 and Ishikura et al. 2008) Medium (Cosemans et al. 2007)	Medium (short BL)	Very high (Cosemans et al. 2007)	High
WR after RD (Pilo et al. 2007; Cosemans et al. 2009 and Yoshimoto et al. 2011)	Medium	Very high (Writeback)	Very high (full BL swing)	High
Pseudo 8T (Sharma et al. 2010)	Low (sequential RD operation + low swing GBL)	High (short BL + VDD/2 pre-charge)	Low (low swing GBL + VDD/2 LBL + charge recycling on LBL)	High
HBSBL (Sharma et al. 2011)	Very high (segment buffers for LBL + pre-charged read bufferforGBL)	Very high (short BL + VDD/2 pre charge + parasitic cap isolation)	Medium (low swing GBL + VDD/2 LBL)	High
Cascaded BL + WR after RD (kushida et al. 2009)	Medium/Low (sequential operation)	Very High (parasitic cap isolation + Write-back)	Very High (full swing GBL + write back)	Medium (very high density)
Litho optimized BL (Sharma et al. 2012)	High (pre-charged read buffer for GBL)	Medium (short BL)	High/medium (low swing BL)	Low

With the result area of our local assist circuitry is 31 % less compared with the conventional local assist circuitry.

Table 4.1. summarizes different local bit-line architectures. The local assist circuit techniques proposed here solves the issues associated with the increased device variations at the scaled voltage levels for the advance sub-nanometric technologies. Hierarchical divided bit-line improves cell stability, performance, and the energy consumption at the increased area overhead.

4.8 Conclusion

The Pseudo 8T local architecture (Sharma et.al. 2010) offers an enhanced SRAM cell stability because of the VDD/2 pre-charged local bit-lines for low and medium performance applications. It includes a local sense amplifier on the short local bit-lines and a gated read buffer. The local sense amplifier reduces the impact of the cell read current on access speed, which allows minimum sized high Vt cell transistors, reducing leakage.

The HBS bit-line (Sharma et al. 2011) provides an energy efficient and high performance interface with very high cell stability without resorting to energy expensive WRITE after READ mechanism. It solves the issues associated with SRAM design in the advance sub-nanometric technologies viz. access time degradation and increased power consumption. The energy consumption is reduced by VDD/2 pre-charged bit-lines, possible because of the segmentation done by the segment buffers. The access speed is increased by the use of segment buffers, driving the pre-charged read buffers.

The litho optimized local architecture (Sharma et al. 2012) reduces the transistor count and timing complexity associated with the conventional local assist circuitry. Reduced timing complexity and transistor sizes reduce the energy consumption compared to the conventional hierarchical divided bit-line architectures. The area overhead of this solution is only 9 % compared to 38 % with the existing solutions. The physical regularity in the layout of the local assist circuitry permits the litho optimization thereby eliminating the memory matrix subarray design complexity associated with the placement of logic circuits. Thus the proposed circuit techniques promise the best area-energy-performance optimization compared to the existing solutions.

References

L. Chang et al., An 8T-SRAM for variability tolerance and low-voltage operation in high-performance caches. IEEE J. Solid-State Circ. **43**(4), 956–962 (2008)

S. Cosemans et al., A low power embedded SRAM for wireless applications. IEEE J. Solid-State Circ. **42**(7), 1607–1617 (2007)

S. Cosemans et al., A 3.6pJ/access 480 MHz, 128Kbit on-chip SRAM with 850 MHz boost mode in 90 nm CMOS with tunable sense amplifiers. IEEE J. Solid-State Circ. **44**(7), 2065–2077 (2009)

S. Ishikura et al., A 45 nm 2-port 8T-SRAM using hierarchical replica bit line technique with immunity from simultaneous R/W access issues. IEEE J. Solid-State Circ. **43**(4), 938–944 (2008)

A. Karandikar, K.K. Parhi, Low power SRAM design using hierarchical divided bit-line approach. in *Proceedings. International Conference on Computer Design: VLSI in Computers and Processors*, pp 82–88, (1998)

A. Kawasumi et al., A single-power-supply 0.7 V 1 GHz 45 nm SRAM with an asymmetrical unit- β-ratio memory cell. ISSCC Dig. Tech. Pap. **622**, 382–383 (2008)

K. Kushida et al., A 0.7 V single-supply SRAM With 0.495 μm^2 cell in 65 nm technology utilizing self-write-back sense amplifier and cascaded bit line scheme. IEEE J. Solid-State Circ. **44**(4), 1192–1198 (2009)

H. Pilo et al., An SRAM design in 65 nm technology node featuring read and write-assist circuits to expand operating voltage. IEEE J. Solid-State Circ. **42**(4), 813–819 (2007)

V. Sharma et.al., Hierarchical buffered segmented bit-lines. *US Patent no. 13/105,806*, Nov 2011

V. Sharma et al., *Ultra Low Power Litho Friendly Local Assist Circuitry For Variability Resilient 8T SRAM Cell, Design Automation and Test in Europe (DATE)* (Dresden, March, 2012), pp. 11–17

V. Sharma, et.al., A 4.4pJ/Access 80 MHz, 2 K Word X 64b memory with write masking feature and variability resilient multi-sized sense amplifier redundancy for W.S Nodes. *Proceedings of IEEE European Solid State Circuits Conference (ESSCIRC)*, pp. 358–361, Sept 2010

B.D. Yang, L.S. Kim, A low-power SRAM using hierarchical bit line and local sense amplifiers. IEEE J. Solid-State Circ. **40**(6), 1366–1376 (2005)

S.Yoshimoto et.al., A 40 nm 0.5 V 20.1 uW/MHz 8T SRAM with low-energy disturb mitigation scheme *Symposium on VLSI Circuits Digest of Technical Papers*, pp. 72–73, June 2011

Chapter 5
SRAM Energy Reduction Techniques

The memories are the most vulnerable to ever increasing process variations in advanced technology nodes as discussed in previous chapters. SRAM bit cell functional parameter degradation due to increasing variability and decreasing power supply is of utmost concern. The increasing intra die variations degrade cell read current, read SNM, and the write margin of the SRAM cell. The SRAM cell design, voltage optimization, and circuit design techniques are required which enhances the operating margin of SRAM and also reduces the energy consumption. The previous chapters discuss only about enhancing the operating margins of SRAM. The energy consumption perspective was not covered. This chapter discusses about various circuit design techniques which result in ultra low energy SRAM operation.

5.1 SRAM Array Leakage Reduction

Technology scaling has enabled the number of transistors in microprocessors to more than double for every scaling technology node. But at the same time leakage current has also increased. The SRAM array leakage current is a major source of static energy consumption for the microprocessors. Static energy consumption of SRAM scales directly with the array size. In order to realize high area efficient SRAM design larger array size is recommended (Bhattacharya et al. 2008). With the result SRAM array leakage contribution in the energy consumption increases. The increased leakage current not only increases the static energy consumption but also degraded the performance of SRAM. The critical delay in SRAM is limited by the required bit-line discharge for the accessed SRAM cell.

The degraded read current and increased bit cell leakage current spoils the I_{ON}/I_{OFF} and increases the delay and the probability of the read access failures. Figure 5.1 shows read current versus the leakage current for the different voltage

V. Sharma et al., *SRAM Design for Wireless Sensor Networks*,
Analog Circuits and Signal Processing, DOI: 10.1007/978-1-4614-4039-0_5,

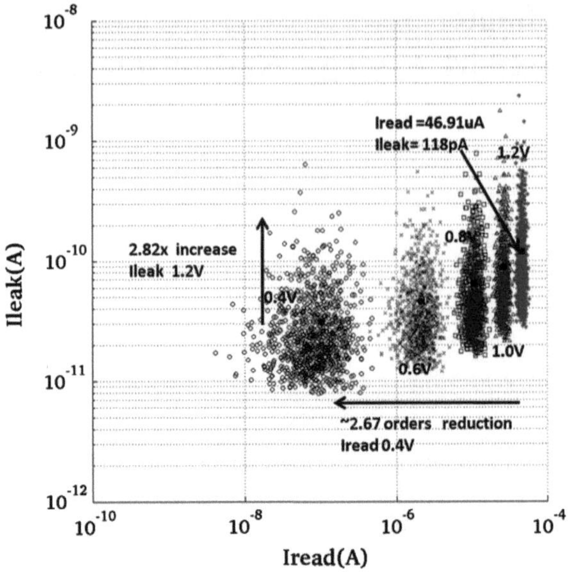

Fig. 5.1 SRAM cell read current versus leakage in 65 nm

levels. Clearly, for the scaled voltage levels the SRAM cell read current and the leakage current decreases. The SRAM array leakage minimization has become vital in realizing ultra low energy variability resilient SRAM designs.

Use of the HVT devices for the SRAM cell mitigates the additional leakage but also results in the lower gate drive thereby reducing the read current. The degradation in read current is further aggravated by the ever increasing process variations. The time required for the development of the bit-line voltage difference increases. This severely increases the read access time, impacting performance (Fig. 5.2). There are many circuit techniques proposed to mitigate the increased leakage without degradation of read current.

The SRAM array leakage minimization techniques can be broadly categorized into two categories:

1. Leakage compensation based techniques: These techniques first detect the bit-line leakage and then compensate for it.
2. Leakage cutoff based techniques: These techniques eliminate the leakage current.

5.1.1 Leakage Compensation-Based Techniques

The leakage compensation-based techniques target the signal integrity issues and requires the injection of the static current to counteract the signal degradation. Various leakage compensation-based techniques are as follows:

Agawa et al. (2001) proposes the bit-line leakage compensation technique that first detects the bit-line leakage and then injects a same magnitude of current into

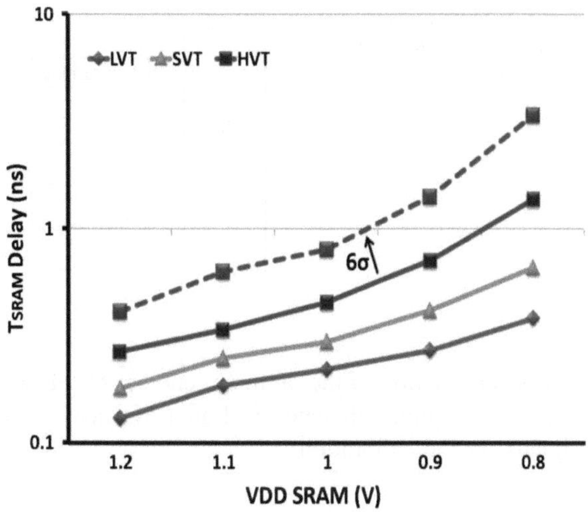

Fig. 5.2 Mean access versus VDD for different Vt of read buffer 8T SRAM. Column height of 256 SRAM cells, nominal process corner, 65 nm LP, and T_{SRAM} delay defined as the time taken to develop 100 mV of RBL swing after the word line assertion

the bit-line. The bit-line leakage for the accessed memory matrix column is detected with a PMOS diode during the precharge phase. The detected voltage drop due to bit-line leakage is stored in a capacitor. Then this voltage drop information is used to inject the same magnitude of current into the bit-line during the SRAM cell assertion phase. This technique results in an overall increase in the static energy because of the injection of leakage current. The dynamic current mirror-based analog circuit used in Agawa et al. (2001) is highly susceptible to the process variations. Therefore, its effectiveness to compensate bit-line leakage is limited for the advanced subnanometric nodes.

Marginal bit-line leakage compensation scheme (Kim et al. 2009) injects a marginal compensation current and compensates for the read bit-line leakage of the un-accessed SRAM cells. It uses a replica bit-line structure. The leakage current information of the fixed data pattern stored on this replica bit-line is compensated by turning ON the compensation devices. The same control information used for the compensation devices of the replica structure is used for enabling the compensation devices of the accessed SRAM columns. This is how the bit-line leakage is compensated. However, the data dependency of the bit-line leakage compensation current generated by the replica structure deteriorates; the sensing margin for the accessed matrix columns having different data patterns compared to the replica bit-line column. The replica-based structure for the marginal compensation current generation is not effective against intra die variations and also the body biasing of the marginal compensation devices used to address this issue is less effective for the advance technology nodes.

X-Calibration technique (Lai and Huang 2008) transforms the bit-line leakage into an offset voltage and then cancels this offset voltage with the help of the calibration circuitry. The calibration circuitry consists of a crossing structure of the transistor switches and capacitor. The effectiveness of X-Calibration technique is dependent on the coupling capacitors used for the offset voltage cancellation.

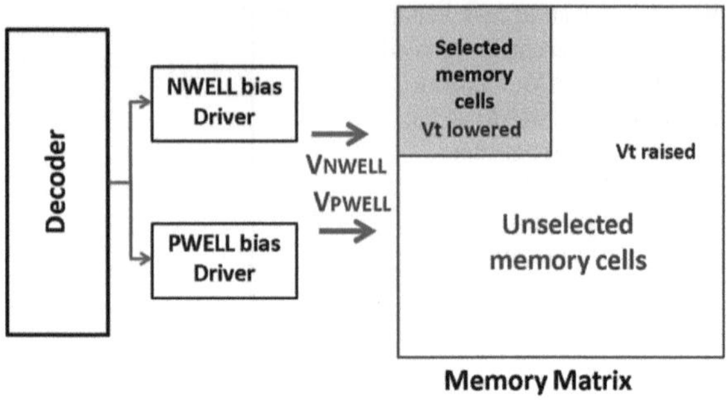

Fig. 5.3 Dynamic leakage cut-off scheme (Kawaguchi et.al. 1998)

The determination of the optimum choice for the coupling capacitors is a tradeoff between coupling efficiency and area overhead. It also requires complex control signals for the calibration.

5.1.2 Leakage CutOff Based Techniques

The leakage cutoff based techniques does not involve any injection of compensation current and are more suitable for realizing low energy SRAMs. Various leakage cutoff based techniques are as follows:

5.1.2.1 Dynamic Leakage CutOff Scheme

Kawaguchi et al. (1998) proposes a dynamic leakage cutoff scheme which applies dynamic bias voltage on the n-well and p-well of the memory cells. Figure 5.3 shows the concept of dynamic leakage cutoff scheme. It ensures high access speed by lowering the Vt of the selected memory cells and subthreshold leakage is mitigated by increasing the Vt for the unselected memory cells with the dynamic adjustment of the well bias voltage. This method of reducing the bit-line leakage has an increased delay penalty associated with the generation of reverse substrate bias, also there is an energy overhead associated with biasing the substrate of a SRAM array. Additionally, the impact of body bias in controlling the threshold voltage is limited for the advance subnanometric technology nodes. Figure 5.4 shows the SRAM array leakage without n-well bias versus with 0.1 V of n-well bias applied for 65 nm LP technology node. Only 5 % reduction in SRAM array leakage is achieved with 0.1 V of applied bias.

5.1.2.2 Gated Read Buffer Based Local Architecture (Pseudo 8T)

Sharma et al. (2010) utilizes HVT transistor based SRAM array. Figure 5.5 shows an accessed 6T SRAM cell and the gated read buffer used in the local bit-line

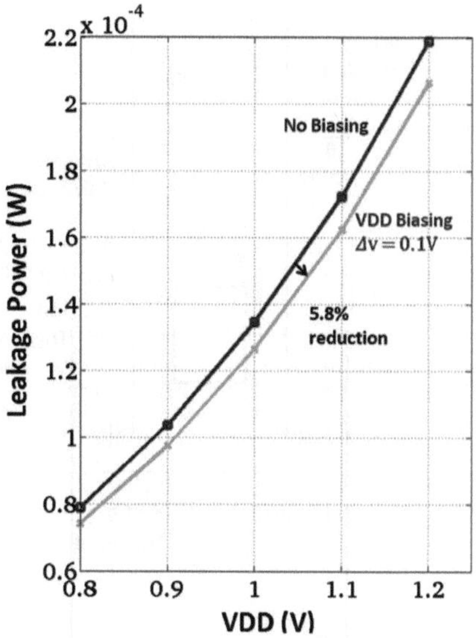

Fig. 5.4 Limiting impact of nwell biasing for leakage reduction for the advance technology nodes. SRAM array consisting of 512 × 512 cells. Nominal process corner, Temp = 25 °C, 65 nm LP and biasing voltage used is 0.1 V

architecture. The required I_{Access} is delivered by an upsized gated read buffer. This buffer is enabled, only for a limited period during the READ operation thereby reducing the local bit-line leakage. The leakage current of the VDD/2 precharged local bit-lines(LBLs) is not that high because of the gated read buffer and the use of HVT transistors for SRAM cells. The leakage current of VDD/2 precharged LBLs constitutes 9 % of the total measured leakage current.

The accessed SRAM cell creates a small voltage swing Δv on the VDD/2 precharged short LBLs. The voltage swing Δv is amplified to full swing voltage level by the local sense amplifier. The read buffer is enabled only during the READ operation. The VDD side of the local bit-line trips the NMOS access transistor of the activated read buffer. The read buffer delivers I_{Access}. The voltage swing, (50 mV < ΔV < 100 mV) on the high capacitive vertical global bit-lines created by I_{Access} of the activated read buffer predominantly determines the memory access time. This is how the access time dependence on the cell read current is relaxed. Then the SRAM cell is sized in favor of improving the cell stability. The read buffer activated for a short duration during READ operation results in a pseudo 8T SRAM cell type READ operation. The gated read buffer provides differential read sensing thereby eliminating the issues associated with the single ended sensing of a conventional read buffered 8T SRAM cell.

Impact on active leakage: There are no current paths for the unselected local bit slices as the NMOS access transistors remain in the cutoff region. For the unselected local bit slices, the gate terminals of NMOS access transistors are biased at the precharged voltage of the short LBLs (VDD/2). The source terminals of NMOS access transistors of the pulled down global bit-line are at VDD/4 − ΔVgbl. Therefore, the gate to source voltage of the unselected local bit slices NMOS access transistors (V_{GS} = VDD/4 + ΔVgbl, 0.2 V + ΔVgbl) is smaller

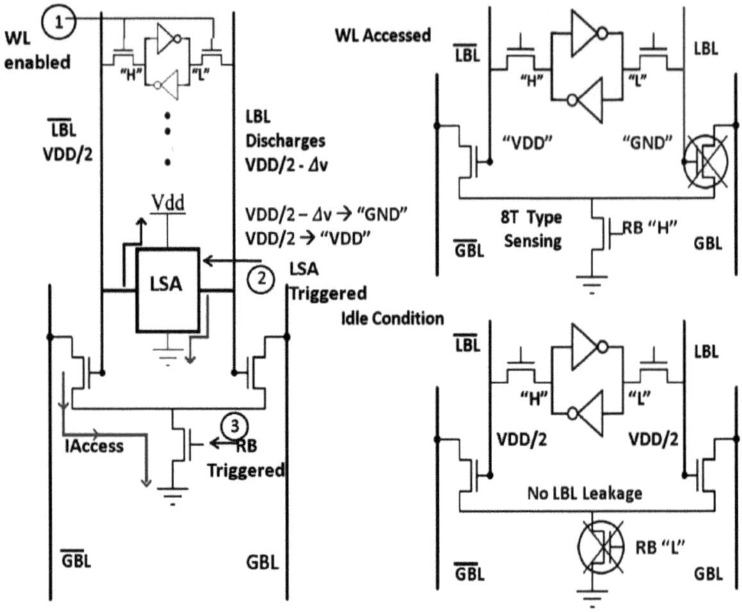

Fig. 5.5 Gated read buffer [Pseudo 8T]: concept (Sharma et al. 2010)

than the threshold voltage (Vtn, HVT transistors). Since the gated read buffer is shared with all the 6T SRAM cells in the local hierarchy. The area penalty normally associated with the use of a conventional read buffered 8T SRAM cell and with improved 8T SRAM cells as discussed in Chap. 2 enabling differential read sensing is minimized.

5.1.2.3 BVSS Driver: Peripheral Assist Technique for 8T SRAM Array

Verma et al. (2009) proposed a BVSS driver, a peripheral assist to limit the extra bit-line leakage associated with read decoupled 8T SRAM cell. The footers (BVSS) of two stack NMOS transistors of un-accessed 8T SRAM cells are pulled high (VDD) with the assistance of BVSS drivers (Fig. 5.6). The voltage drop across the stacked NMOS transistors of 8T SRAM cell becomes approximately 0 V and the read bit-line leakage reduces. The design also uses a charge pump circuit to boost the performance of the BVSS driver at the reduced voltage levels. However, there is a potential risk of severe read current degradation due to the IR drop. The drive current of accessed memory cells from all the columns flow through the BVSS node, resulting in a huge IR drop. The gate-to-source voltage (*V*gs) of the read buffers get reduced thereby degrading read current. Sinangil et al. (2009) addresses this problem by reducing the resistance of BVSS node with the layout optimization.

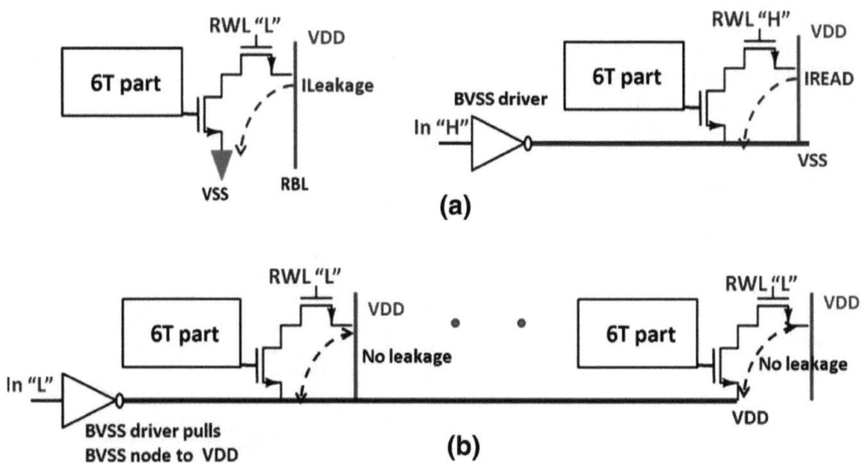

Fig. 5.6 BVSS driver: concept (Verma et al. 2009)

This technique requires switching of the highly capacitive BVSS node of an accessed memory word from VDD to VSS resulting in an excessive energy consumption penalty. This energy overhead is more critical for the above threshold voltage range. Figure 5.7 shows the switching energy consumption associated with the BVSS driver implementation for the wide voltage range. For example, switching energy can be as high as 0.32 pJ for the word length of 128 cells. This energy overhead cannot be neglected for the SRAM design targeting 2.5 pJ/access or even less energy consumption.

5.1.2.4 RSDVt 8T SRAM Array

The reduced swing dual Vt 8T SRAM cell (Fig. 5.8) proposed in this work offers a simple solution which reduces the static leakage current for the target performance level without any energy and area overhead (Sharma et.al. 2011). The write port (6T structure, Fig. 5.8) consists of HVT transistors. The write port will have a marginal impact on read access time. But it results in an overwhelming reduction in the leakage current, as the cross-coupled inverters contribute to 70 % of the leakage. However, using HVT 6T structure degrades the write ability and the WRITE access time. The reduced write ability is improved by using Mimicked Negative Bit-line technique as discussed in Chap. 3. The read access time is always more critical compared to write access time, so not an issue as long it meets target performance requirement.

The two stacked NMOS transistors (read buffer) determine the read access time. Therefore, the read port consists of SVT transistors instead of HVT for the target performance requirement. Figure 5.8 shows leakage current versus read current for the different Vt devices for 8T SRAM cell.

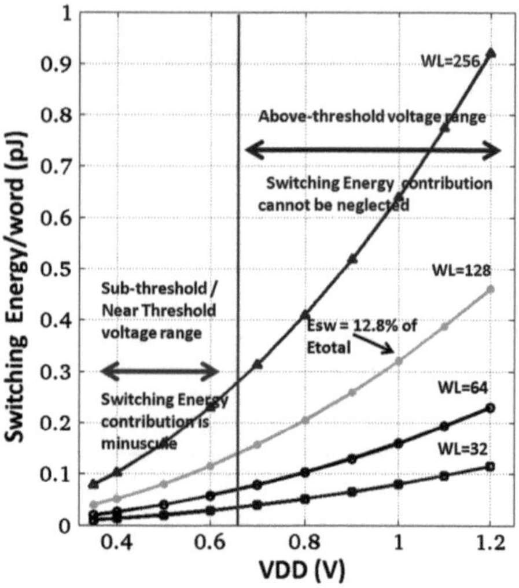

Fig. 5.7 Switching energy consumption associated with BVSS driver implementation for different word lengths ranging from 32 to 256 bits. Energy cost associated with switching of read buffer VSS of 8T SRAM cell cannot be neglected for realizing ultra low energy above threshold SRAM. Nominal process corner, Temp = 25 °C, 65 nm LP

The dual Vt 8T SRAM cell standby leakage at 0.8 V is 47.7 pA/cell which is approximately half of the single Vt 8T SRAM cell at VDD = 0.8 V for the same read current of 16.2 uA/cell. The standby leakage of dual Vt 8T SRAM cell is further reduced by reducing the read bit-line precharge voltage. The SRAM cell stability degradation due to the reduction of bit-line precharge voltage is not an issue with 8T cell because the internal storage nodes are isolated from the read bit-line.

The read bit-line leakage is a state dependent leakage component in 8T SRAM cell, which contributes approximately 30 % leakage current for the worst data pattern (Q = "H"). Figure 5.9 shows an impact of bit-line configuration (precharge value and state) on the normalized leakage current for a dual Vt 8T SRAM cell. The 8T SRAM cell can rely on the static write bit-lines because of the isolation of the write port from the read port. The static write bit-lines help in reducing the leakage consumption. Furthermore, 0.2 V precharge voltage on the read bit-line results in 3.5x reduction in leakage current for the worst data pattern (Q = "H"). Dual Vt 8T SRAM cell with 0.2 V read bit-line precharge voltage consumes only 20 % more leakage current compared to the VDD precharged RBL single Vt HVT 8T SRAM cell and delivers 45 % more read current.

5.2 Dynamic WRITE Energy Reduction

During READ operation, the voltage swing of the bit-lines is limited to a smaller value (depending on the resolution of the SA employed), whereas the WRITE operation requires full voltage swing on the bit-lines. The charging and discharging of the high capacitive bit-lines account for the major proportion of the dynamic energy consumption. This results in energy consumption of the WRITE

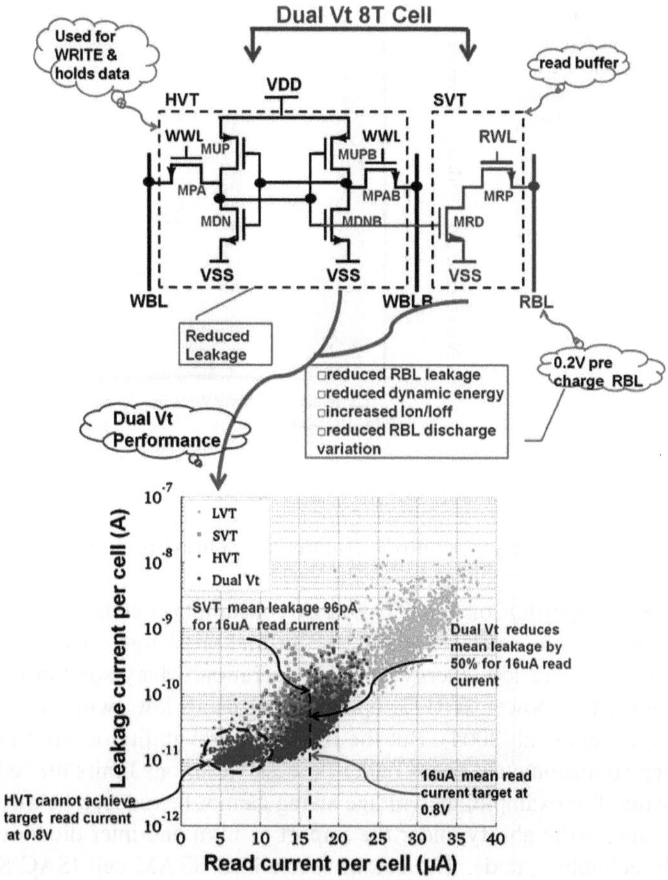

Fig. 5.8 Reduced swing dual Vt 8T SRAM cell (Sharma et al. 2011)

operation to be much higher compared to the READ operation. The WRITE operation dynamic energy consumption is a critical parameter in the design of the ultra low energy SRAMs. This section will outline various circuit techniques for realizing low energy WRITE operations.

Figure 5.10 shows WRITE energy consumption per bit versus VDD of SRAM array for 40 nm and 65 nm LP technology. Clearly, reducing the VDD of SRAM array helps in reducing the WRITE energy consumption. But the degraded write ability at reduced supply voltage level necessitates the use of WM improvement technique. The WRITE energy reduction techniques based on low VDD SRAM have to use write assists for improving the degraded WM at the scaled voltage levels. For example, voltage optimization techniques as discussed in Chap. 3. The timing control circuit techniques can also be used for reducing the energy consumption as proposed by Yamaoka et al. 2006. But the limited energy reduction gain and limited effectiveness against intra die variation limits the

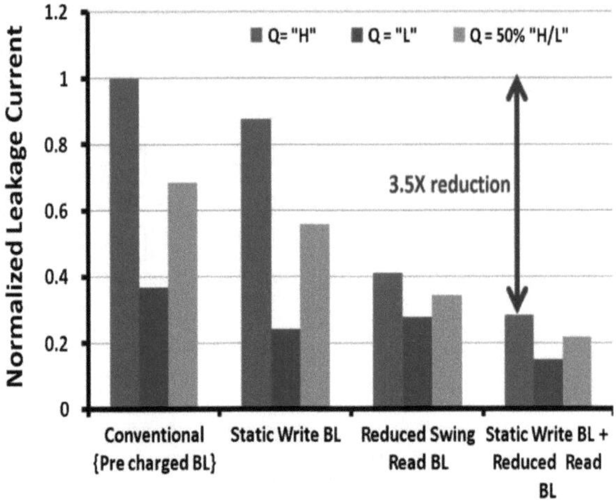

Fig. 5.9 Normalized leakage current of dual Vt 8T SRAM cell (HVT latch and SVT RB). VDD = 0.8 V, reduced read BL = 0.2 V for the nominal process corner

applicability of the timing control circuits in achieving low energy WRITE operation for advanced technology node.

The major proportion of WRITE energy consumption consists of the bit-line energy. Therefore, the selective scaling of bit-line precharge voltage holds a key for achieving an ultra low energy WRITE operation. Many solutions have been proposed for a low power WRITE operation by using low swing datalines (Mai et.al. 1998; Kanda et.al. 2004). But the increased probability of write failures for the advance subnanometric nodes under process variations limits the reduction of voltage swing. For example, the voltage swing cannot be reduced below 0.45 V in order to ensure write ability under the impact of intra and inter die variations for 90 nm LP technology node. The use of specialized SRAM cell (SAC-SRAM—a low power sense amplifying cell) can enable further scaling of bit-line precharge voltage at the expense of an increased area overhead.

Similarly, charge recycling (CR-SRAM) (Kim et al. 2008) reduces the voltage swing on the bit-lines during write cycles by recycling the charge from the neighboring bit-line capacitances. The write ability degradation due to the charge loss because of leakage and the power/delay penalty for the write-start up overshadows the WRITE energy reduction. The WRITE operation with low swing data transfers from high capacitive global bit-lines onto much less capacitive LBLs, whereas the full swing conversion is done by the local write receiver (WR) results in much higher energy reduction gain. The energy reduction gain is further improved and the area overhead associated with the use of a local WR is remedied by Sharma et al. (2010). The issues related with complex memory matrix optimization and the timing control is remedied by litho optimized low swing static WRITE operation Sharma et al. (2012). This achieves the maximum energy reduction gain and also rules out complex memory array optimization. The implementation details of various low WRITE energy techniques are as follows:-

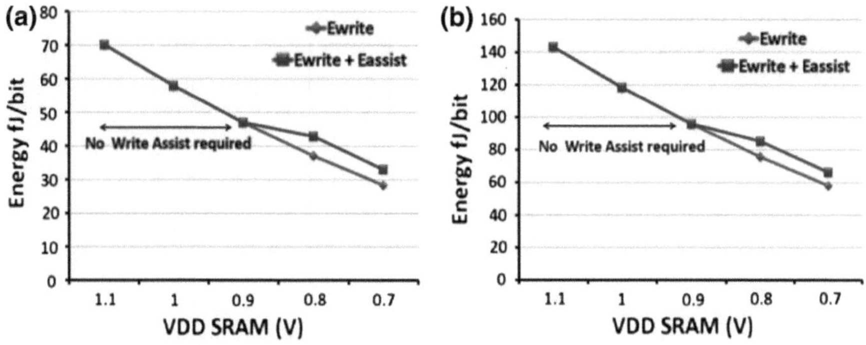

Fig. 5.10 Energy consumption per bit versus VDD of SRAM array. Column height of 256 SRAM cells, nominal process corner and T = 25 °C. **a** 40 nm LP. **b** 65 nm LP. Eassit includes only the energy consumption associated with WL modulation

5.2.1 Write Replica Circuit for Low Power Operation

During the WRITE operation, the BLs in selected column are switched full signal and the bit-lines for the unselected column are in the pseudo READ state (half select condition) (Yamaoka et al. 2006). The word line duration should be long enough to ensure writeability of the accessed SRAM cells. But at the same time the WL duration is directly proportional to the unnecessary energy consumed for the un-accessed bit-lines because of the pseudo READ operation. This problem is remedied by using WRITE replica circuit Yamaoka et al. (2006) Fig. 5.11. During the WRITE operation, the WR_EN signal is triggered and the write replica transistor (MWR_Tr) is enabled and the replica bit-line RBL discharges. The discharging RBL signal combines with the decoder control signal and disables the asserted WL signal.

The NMOS MWR_Tr adjusts the timing for different condition of threshold voltage. When the threshold voltage is low, the WRITE operation of the SRAM cell is fast and so is the discharge process of RBL. The WL duration is shortened and the excessive bit-line discharge is prevented. Similarly, for the high threshold voltage the RBL discharging takes longer. The WL duration is increased in order to ensure the writeability for the accessed SRAM cells. This results in 18 % lower energy consumption for the WRITE operation. However, this technique does not reduce the energy consumption associated with the full swing bit-lines for the accessed cell. Therefore, the energy reduction benefits with WRITE replica circuit are limited.

5.2.2 Charge Recycling SRAM

Low power SRAM using charge recycling (CR-SRAM) reduces the voltage swing on the bit-lines during write cycles (Kim et al. 2008). Low voltage swing for each bit-line is obtained by the recycled charge from the neighboring bit-line capacitance.

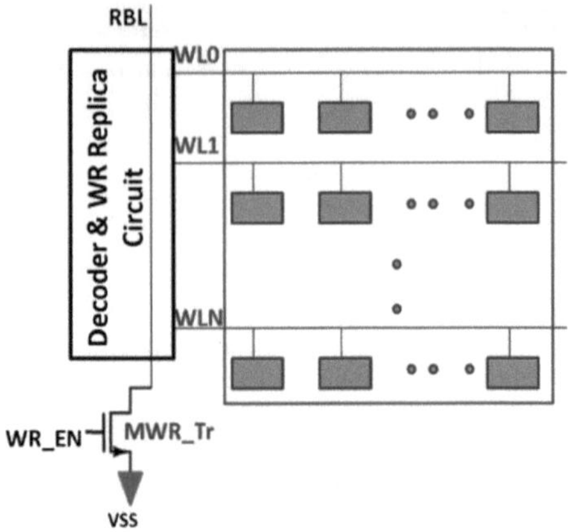

Fig. 5.11 Circuit diagram of WRITE replica circuit Yamaoka et al. (2006)

The WRITE operation is performed in the equalization and the evaluation modes. In equalization mode, two bit-lines in a bit-line are precharged to a common voltage. This common voltage is established by consecutive charge—recycling WRITE operations. In evaluation mode, two bit-lines in a bit-line pair have different voltages.

!The degradation in writeability due to the charge loss because of the bit-line leakage is a limiting factor. CR-SRAM proposes to reduce the bit-line leakage by increasing the source line voltage. It requires all the bit-lines to be precharged back to VDD during READ operation. This results in power and delay overheads for the write-start up. In addition, increased leakage and the reduced impact of body bias in advanced technology nodes makes it a less optimum choice for realizing low energy SRAMs.

5.2.3 Sense Amplifying SRAM Cell (SAC-SRAM)

Reduced voltage swing on the datalines is an efficient method to reduce the dynamic energy consumption during WRITE operation (Kanda et al. 2004). The increased probability of write failures for the advance subnanometric nodes under process variations limits the reduction of voltage swing. The techniques relying on the selectively lowering of the bit-line voltage during write cycles as reported in (Mai et al. 1998) is becoming less effective for realizing an ultra low energy WRITE operation. This problem is remedied by Kanda et al. (2004).

SAC-SRAM is a low power sense amplifying cell, which enables the use of the small swing bit-lines during the WRITE operation. An additional NMOS transistor is connected to the source of driver NMOS transistors (VSS switch) in a SRAM 6T cell. The VSS switch in SAC is turned OFF before the WL reaches VDD (asserted). In other words, the SRAM cells to be written have no connection to VSS (floating). The floating VSS for the accessed SRAM cell enhances the write ability (Chap. 3, voltage optimization techniques reducing the strength of the latch structure). With the result only small bit-line voltage difference suffices the WR operation. But the SAC-SRAM suffers from an increased area overhead and the reduced static noise margins because of the ground voltage connection of a SRAM cell via NMOS transistor.

5.2.4 Low Swing WRITE Operation

This low power WRITE operation uses hierarchical bit-lines and a local WR (Yang and Kim 2005, Cosemans et al. 2007). The WRITE operation is executed with low swing data transfers from high capacitive global bit-lines onto much less capacitive LBLs where the full swing conversion is done by the local WR. Cosemans et al. (2007) reduces the number of local WR as a WR is shared between two local blocks. This necessitates the use of a dedicated write bit-line. The WRITE operation proposed features low swing data transfer on the high capacitive global bit-lines with local amplification on the short LBLs. The local WRITE operation at the local bit slice level and the associated timing waveforms are shown in Fig. 5.12. The write data transfer (WDT) signal transfers the data from the global bit-lines onto the short LBLs. After the data transfering from the global bit-lines, the local WR is activated (LWR_en). The enabled local WR resolves the low swing data information $\Delta v_{\mathrm{WRITE}}$ to the full swing voltage level on much less capacitive short LBLs. There is no write-ability degradation with the use of low swing data information on the global bit-lines, since the LBLs of the accessed SRAM cell are full swing. The energy consumption increment due to full swing voltage is much less because of the reduction of effective bit-line capacitance with the hierarchical bit-line structure.

5.2.5 Low Swing WRITE with WRITE Masking

Sharma et al. (2010) proposed to reuse the local sense amplifier used during READ operation for resolving Δv during WRITE operation. Therefore, the local sense amplifier acts as a local WR amplifying $\Delta v_{\mathrm{WRITE}}$ to the full swing voltage level on the short LBLs and consumes on 4.7 fJ/decision during WRITE operation. The reuse of local sense amplifier designed for READ operation as a local WR during WRITE operation rules out an extra dedicated WR in each local bit slice. It also proposes a WRITE masking feature.

The WRITE masking acts as an energy control knob feature. This control feature of selectively masking the WRITE operation for certain bits of the data word length has a direct energy reduction implication. This feature enables the operating system to proactively reduce the WRITE energy consumption. The data correlation is exploited and the write operation for the certain set of bits of the data word is prevented.

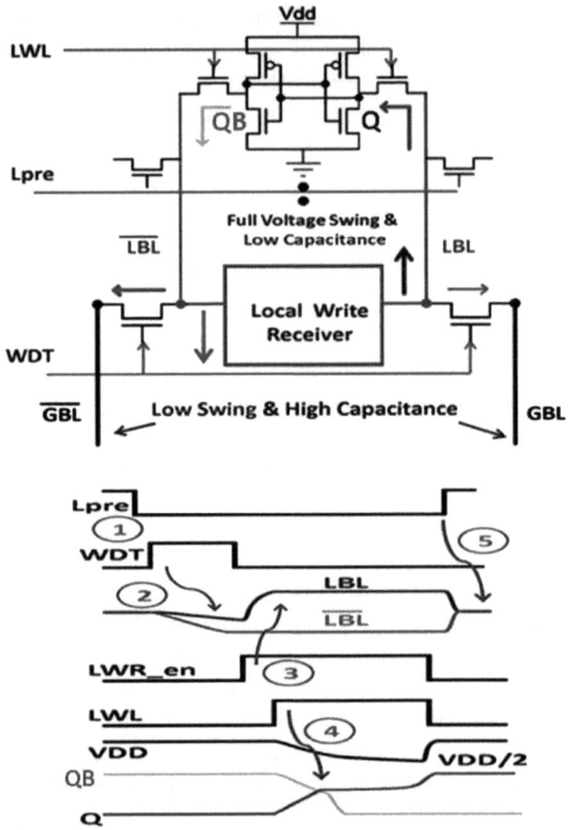

Fig. 5.12 WRITE operation and timing waveform: local sense amplifier used during READ operation acts as a local WR for WRITE operation

In this implementation, WRITE masking is done for a quarter of a word by the mask decode logic used in each word block as shown in Fig. 5.13. The masking bits are decoded at the decoder stage and masks out the events for the WRITE operation. For the quarter of masked word {16 bits} the global bit-lines are not loaded with the data input information. The write multiplexers at the local bit-line architecture are not activated and the LBLs remain precharged at VDD/2. The word line for the masked quarter of cells and the corresponding WR of the masked local bit slices are not activated. The VDD of 16 local bit slices in a word block is shared. The VDD of the SRAM cells for the masked local bit slices is not switched to VDD/2.

The dynamic energy consumption of WRITE operation is reduced with the selective activation of circuits for the unmasked quarters of the word. The WRITE energy is reduced by 44 % when writing a three-fourth masked word compared to the full word. The option of masking feature results in an overall decrease in energy consumption for the applications in which the number of partial writes (**Nw**) are relatively high compared to the number of reads (**Nr**).

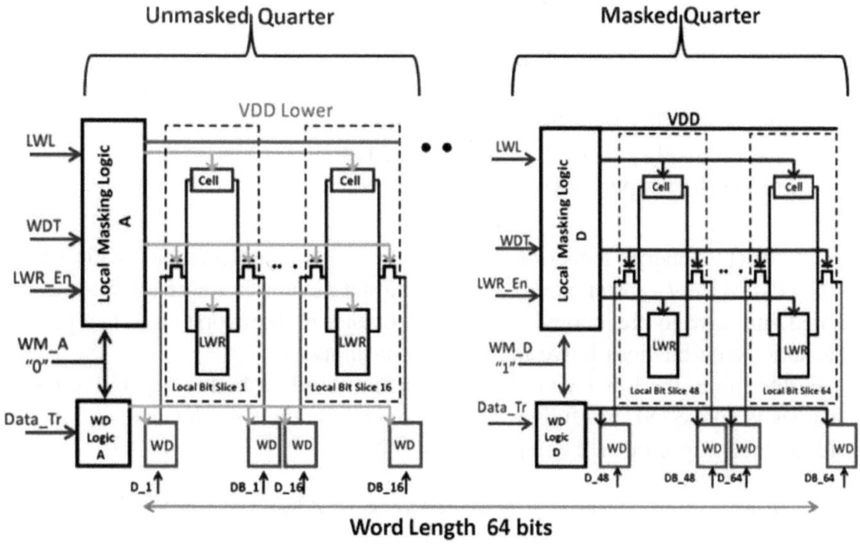

Fig. 5.13 Write masking implementation (Sharma et al. 2010)

The additional decode circuitry required for the selective activation of the assist circuitry for the low energy implementation of the masking feature increases the read energy consumption by 16 %. In applications where the partial write accesses are in good proportion to the read accesses it reduces the overall access energy consumption.

5.2.6 Low Swing Static WRITE operation

In traditional SRAM design, the bit-lines are shared for the READ and WRITE operations (Sharma et al. 2011). The READ and WRITE operations occur in an interleaved fashion. The energy is consumed during each cycle, independent of the data pattern, as the bit-lines must be precharged for the READ operations. The separate read and write bit-lines eliminate the requirement for also precharging the write bit-lines for the READ operation (static write signals). The static write signals help in reducing the WRITE energy consumption. With complementary static signals, energy is only consumed when the data to be written changes from 0 to 1 or from 1 to 0. The worst case WRITE energy consumption per bit associated with the complementary static signals is

$$E_{\text{StaticWBL}} = (P_0 \times P_1 + P_1 \times P_0) \times C_{\text{BL}} \times \text{VDD}^2$$

$$P_1 = 1 - P_0$$
$$= (2 \times P_0 \times (1 - P_0)) \times C_{BL} \times \text{VDD}^2 \quad (5.1)$$

P_0 the probability that "0" must be written
P_1 the probability that a "1" must be written
C_{BL} the bit-line capacitance

The WRITE energy reduction gain with the use of complementary static signals is $2 \times P_0 \times (1 - P_0)$, compared to the traditional memories where the bit-lines are shared and have to precharged (not static). Even with the worst case data pattern, the static write bit-lines halve the energy consumption.

Furthermore, the low swing data information used on highly capacitive global bit-lines is converted into full swing data signals on the short local write bit-lines with a small capacitance C_{LWBL}.

$$E_{\text{LowswingBL}} \cong \left(C_{LWBL} \times \text{VDD} + C_{GWBL} \times \Delta V_{\min,LWR}\right) \times \text{VDD} \quad (5.2)$$

$\Delta V_{\min,LWR}$ required input signal for local WR
C_{LWBL} capacitance of local write bit-line
C_{GWBL} capacitance of global write bit-line

The column height of this design is 256 cells and the number of cells on local hierarchy is 16. Assuming the local write bit-line capacitance with 16 write ports of 8T SRAM cell is 1/16 of the traditional bit-line capacitance.

$$C_{LWBL} \cong {C_{BL}}/{16}$$

$$\Delta V_{\min,LWR} = {\text{VDD}}/{4}$$

Substituting the values in Eq. (5.2)

$$E_{\text{LowswingWBL}} \cong {5}/{16} \times C_{BL} \times \text{VDD}^2 \quad (5.3)$$

The energy consumption per bit for the static low swing write bit-lines is

$$E_{\text{Lowswing,staticWBL}} \cong (2 \times P_0 \times (1 - P_0)) \times {5}/{16} \times C_{BL} \times \text{VDD}^2 \quad (5.4)$$

The local WR used in the local hierarchy with 16 write ports (hierarchical divided write bit-lines) of 8T SRAM cell additionally consumes 6.12 fJ/bit. Clearly, the static write signals for both global and local write bit-lines in addition to low swing signals for global bit-lines with local amplification achieves an extremely low energy WRITE operation. The energy reduction gains with the static low swing write bit-lines is approximately 6.4x compared to the traditional WRITE operation (Fig. 5.14). The WRITE energy consumption for the design implementing LSSWR is 2.5 pJ/access for word length of 64bits at VDD = 0.8 V.

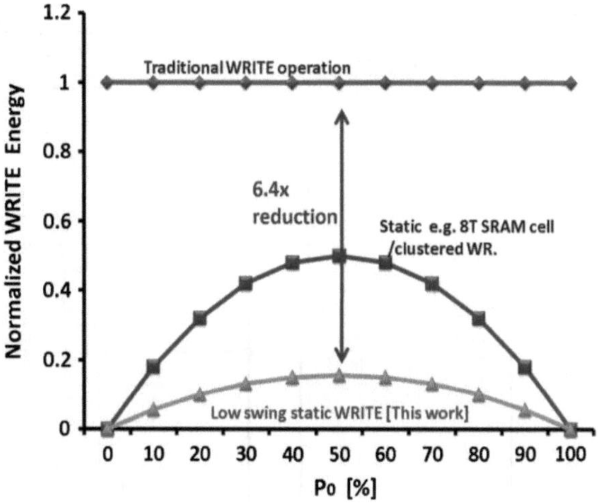

Fig. 5.14 Energy consumption comparison of low swing static WRITE bit-lines with the traditional WRITE operation (precharged complementary bit-lines) and the static complementary bit-lines. It is assumed that P_0 is independent of the position of the bit in the word and in the memory and of the previous value written to the same column

5.2.7 Litho Optimized Low Swing Static WRITE

Sharma et al. (2012) proposed to replace the strobed local WR with two cross-coupled inverters, NS-LWR (Fig. 5.15). This architecture reduces the timing complexity associated with strobe signal generation of the local WR. The WL_WR activation signal for the WR MUX (write pass transistors) transfers the low swing information onto the LBLs and also serves the purpose of triggering the regenerative action of the two cross-coupled inverters. The pulsed WL_WR signal isolates the nodes of the cross-coupled pair from the highly capacitive bit-lines. This architecture also implements the differential VSS biasing technique, which allows the independent tuning of the GND connection of the cross-coupled inverters of SRAM cells and the NS-LWR as discussed in Chap. 3. The low swing input data information is converted to full swing by the regenerative action of the two cross-coupled inverters (NS-LWR) on the short local write bit-lines. The SRAM cell type structure of NS-LWR & WR MUX enables litho optimization of the local assist circuitry as discussed in Chap. 4, saves area overhead and reduced timing complexity helps in further reducing the energy consumption.

5.3 Low Energy READ Operation

Unlike WRITE operation, where the scaling of VDD helps in reducing the energy consumption. Low voltage SRAM does not imply low energy READ operation for SRAM. Figure 5.16 shows the ideal energy consumption (Eread) and under the impact of intra die variations [Eread(variation)] per bit for 40 nm and 65 nm SRAM array of column height of 256 cells. Eread(variation) also accounts for the excessive bit-line discharge caused by the fast SRAM cells (positive Vt shifts) in

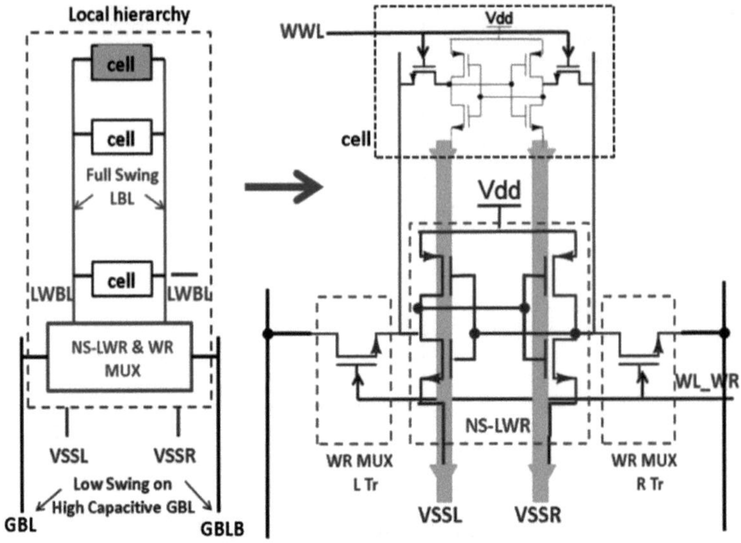

Fig. 5.15 NS-LWR with differential VSS biasing: achieves ultra low energy WRITE operation (only write part of 8T is shown in above figure) and solves the issues related with existing LWR architecture

the time period (word line pulse width) dictated by the slow SRAM cells (negative Vt shifts). The asserted SRAM cell stops discharging the bit-line only when the wordline is deactivated. With the result the average bit-line swing is always larger than ΔV_{min} required by the sense- amplifier. In order to achieve correct operation under all process corners, design margins are taken for the wordline activation. Due to the impact of large intra die variations in subnanometric technologies, different cells of the accessed word on the same die generate different swings with the same wordline activation duration. The swing developed by the slowest cell has to be larger than ΔV_{min} requirement set by the offset mismatches of the sense amplifier. With the result, the average swing of the cells in an accessed word becomes larger than this value, thereby increasing the average energy consumption associated with precharging the high capacitive bit-lines (Fig. 5.17).

The implementation and performance details of various SRAM cells and the local assist circuit techniques are discussed in the previous chapters. Only the energy efficient circuit techniques are discussed here. Various circuit techniques which achieve ultra low energy SRAM READ access are described.

5.3.1 Hierarchical Buffered Bit-lines

The detailed architecture is discussed in previous chapter. Cosemans et al. (2009) proposed low swing global bit-lines. Low swing GBL reduces the energy associated with the precharging of highly capacitive global bit-lines. Low swing GBL also

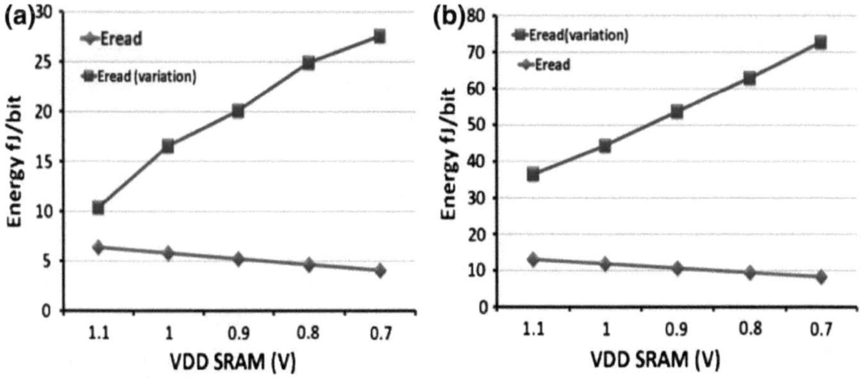

Fig. 5.16 Energy consumption per bit versus VDD of SRAM array. Column height of 256 SRAM cells, nominal process corner and $T = 25$ °C. **a** 40 nm LP. **b** 65 nm LP

limits the impact of process variation in increasing the energy consumption by limiting the swing on GBL. However, there is a major drawback in achieving ultra low energy read access with this approach (Fig. 5.18). Firstly, the low Vt read buffers used increase the leakage power. Secondly, the read buffers used are operating with the full swing precharged LBLs, resulting in an increased dynamic energy consumption.

5.3.2 Pseudo 8T Architecture Based Local Architecture

The local architecture consists of eight high Vt transistor based SRAM cells, local sense amplifier, and the gated read buffer in the local hierarchy (detailed local architecture discussed in previous chapter) (Sharma et al. 2010). The charge recycling involves the equalization of the two bit-lines to VDD/2. Then the datalines are driven to their final values (VDD/VSS). But this transition occurs from the initial state of VDD/2 instead of VDD with the result the energy consumption is halved. In (Sharma et al. 2010), charge recycling occurs because of the toggling of the VDD/2 precharged LBLs by the activation of the local sense amplifier in order to sense the small local bit-line voltage swing created by an accessed SRAM cell. This decreases the energy consumption. Approximately there is a 40 % reduction in the active energy consumption; with charge recycling including the energy cost associated with the activation of the assist circuitry (LSA) compared to a VDD precharged LBLs used in the conventional local bit-line architectures. Figure 5.19 shows pseudo 8T architecture.

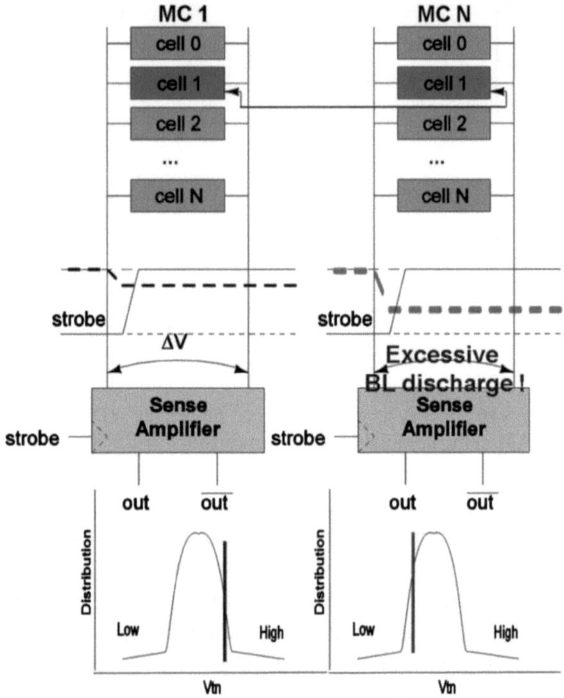

Fig. 5.17 Impact of intra die variations on the READ access energy

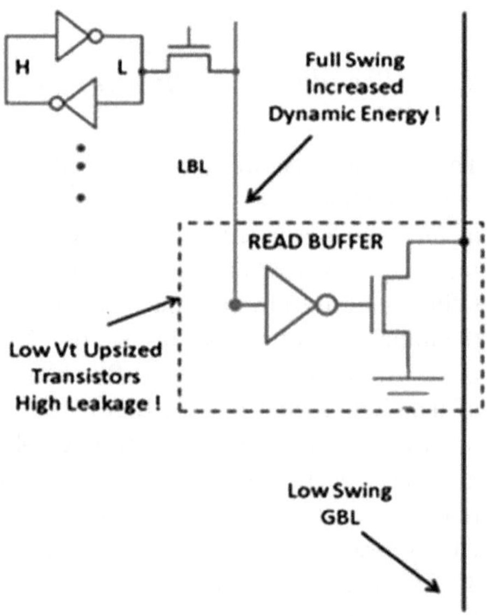

Fig. 5.18 Hierarchical buffered bit-lines Cosemans et al. (2009)

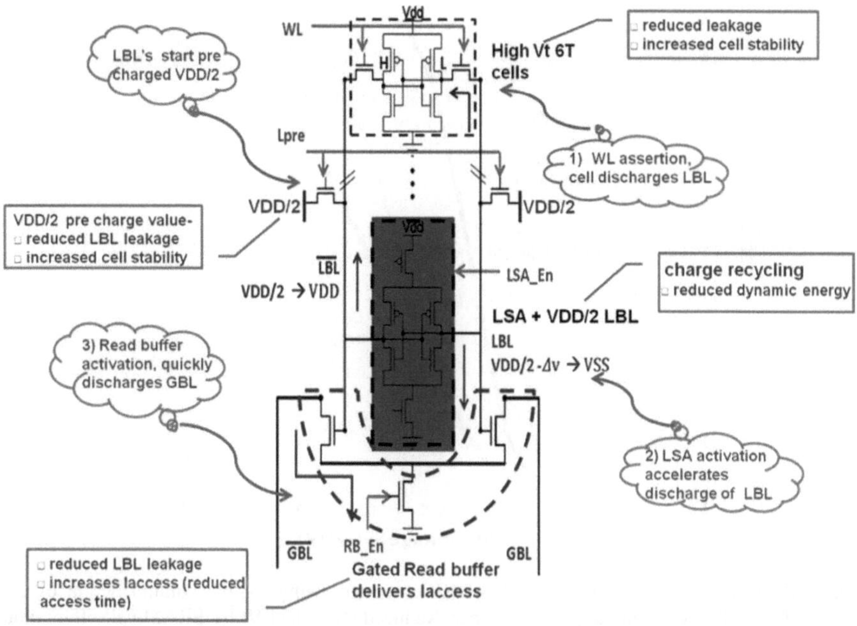

Fig. 5.19 Low energy local assist circuit for 6T cell-based SRAM array (Pseudo 8T architecture). The VDD/2 precharged bit-lines and the local sense amplifier activation in response to accessed SRAM cell results in charge recycling

5.3.3 RSDVt 8T SRAM: Variability Resilient Low Energy Solution

The RSDVt 8T SRAM cell as discussed earlier for leakage reduction also reduces the dynamic energy consumption associated with the read bit-lines (Sharma et al. 2011). The enhanced variability resilience with the reduced read bit-line voltage further mitigates the power consumption associated with unnecessary read bit-line discharge under the impact of intra die variations. Figure 5.20 shows the distribution of RBL voltage swing of 8T SRAM cell at VDD = 0.8 V. The average RBL voltage swing when 99.9 % of the cells have discharged RBL by 100 mV (the target value dictated by the sense amplifier mismatch offset) is 180 mV for the dual Vt 8T SRAM cell with the reduced RBL voltage. Whereas the average discharge when compared to the VDD precharged dual Vt 8T SRAM cell is 270 mV. The spread of the RBL voltage swing with reduced RBL precharged dual Vt 8T SRAM cell is also much smaller, 8 mV compared to 44 mV with VDD precharged dual Vt 8T SRAM cell. This is how, the excessive unnecessary RBL discharging by the faster 8T SRAM cell, in the time duration set for the slowest 8T SRAM cell is avoided.

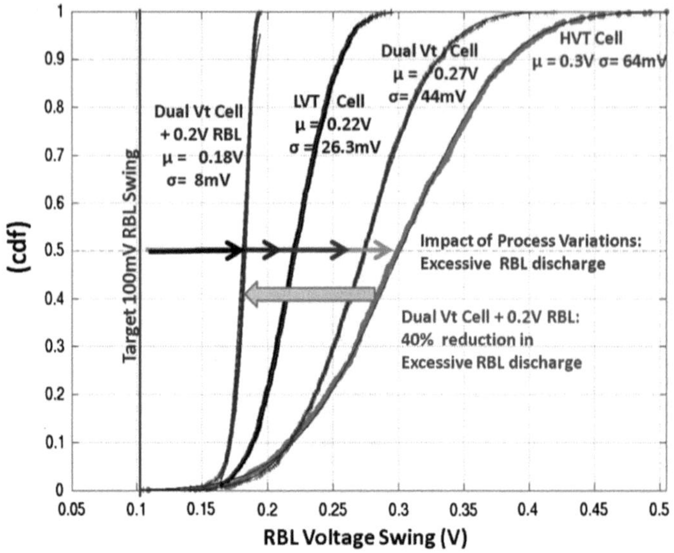

Fig. 5.20 Cumulative distribution (cdf) of RBL voltage swing for the column height of 256 cells, VDD = 0.8 V, Nominal process corner. Reduced RBL dual Vt 8T SRAM cell offers more variability resilience and reduces power consumption associated with bit-lines. SRAM cell assertion pulse width is adjusted to ensure target 100 mV RBL discharge for the slowest 8T SRAM cell

5.4 Compartive Analysis

This section discusses energy consumption associated with various low energy circuit techniques. Figure 5.21 shows SRAM array (512 × 512 cells) leakage current comparison for the array size of 512 × 512 cells, nominal process corner, Temp = 25 °C, 40 nm and 65 nm LP technology. The BVSS Driver (NVer08) achieves the leakage current reduction by pulling the BVSS node to “1.” But it does not optimize for the leakage current contribution from the write port of a 8T SRAM cell. The RSDVt (Sharma et al. 2011) SRAM cell utilizes high Vt write port structure with the floating write bit-line and the reduced swing read bit-line. Therefore, the RSDVt offers the maximum reduction in the leakage current for the 8T SRAM cell-based array. The reduction in the leakage current is in the range of (4.2–4.8)x for 40 nm and 65 nm LP technology.

The Gated RB (Sharma et al. 2010) utilizes the high Vt transistors based SRAM array and also cutoff the bit-line leakage for the un-accessed local bit-line hierarchy. The Gated RB based 6T SRAM array results in (1.9–2.8)x reduction in the leakage reduction for the 6T SRAM cell-based array for 40 nm and 65 nm LP technology.

Figure 5.22 shows WRITE energy consumption for 40 nm LP and 65 nm LP technology. Clearly, LSMWR (Sharma et al. 2010), LSSWR (Sharma et al. 2011), and LLSSWR (Sharma et al. 2012) achieves an ultra low energy WRITE operation

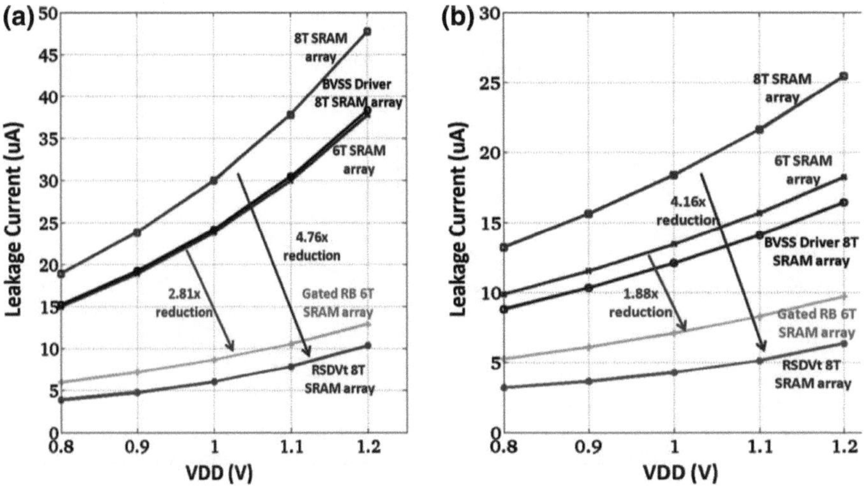

Fig. 5.21 Leakage current comparison for the array size of 512 × 512 cells, nominal process corner, Temp = 25 °C. **a** 40 nm LP. **b** 65 nm LP

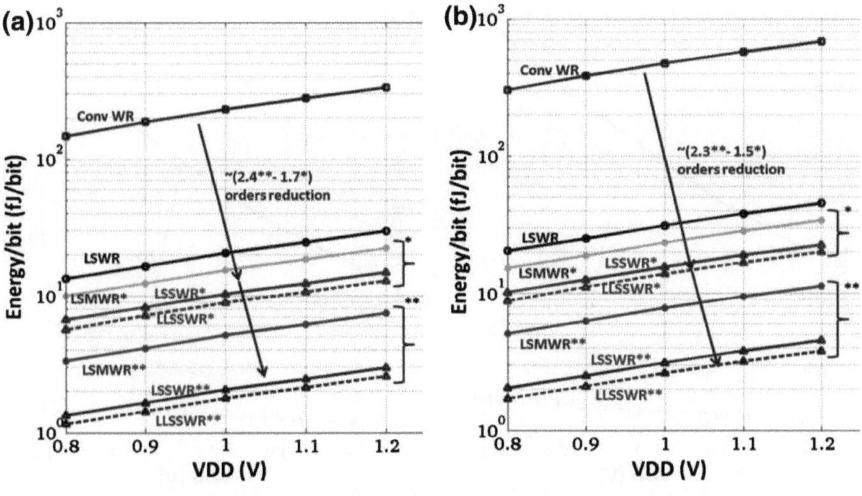

Fig. 5.22 Energy consumption per bit for WRITE operation. Column height = 512 cells, nominal process corner, Temp = 25 °C. **a** 40 nm LP. **b** 65 nm LP *single asterisk*—(¾) un masked quarters for LSMWR, worst case data pattern (50 %, $Prb_{switching}$) for LSSWR. *double asterisk*—(1/4) un masked quarter for LSMWR, data pattern (10 % $Prb_{switching}$) for LSSWR

compared to the conventional WR operation and the low swing write operation (LSWR) (Yang and Kim 2005; Coseman et al. 2007). The litho optimized low swing static write (LLSSWR) (Sharma et al. 2012) offers the maximum reduction in the write energy consumption. This high energy reduction gain is attributed to low swing static signals used on the highly capacitive global bit-lines, reduced bit-line

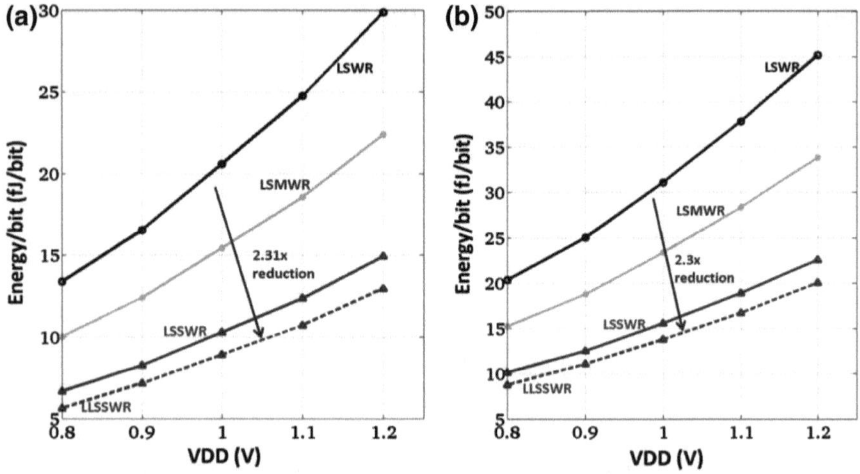

Fig. 5.23 Energy consumption comparison for LSWR. Techniques for the worst case scenario [(3/4) un masked quarters for LSMWR, worst case data pattern (50 %, $Prb_{switching}$) for LSSWR]. Column height = 512 cells, nominal process corner, Temp = 25 °C. **a** 40 nm LP. **b** 65 nm LP

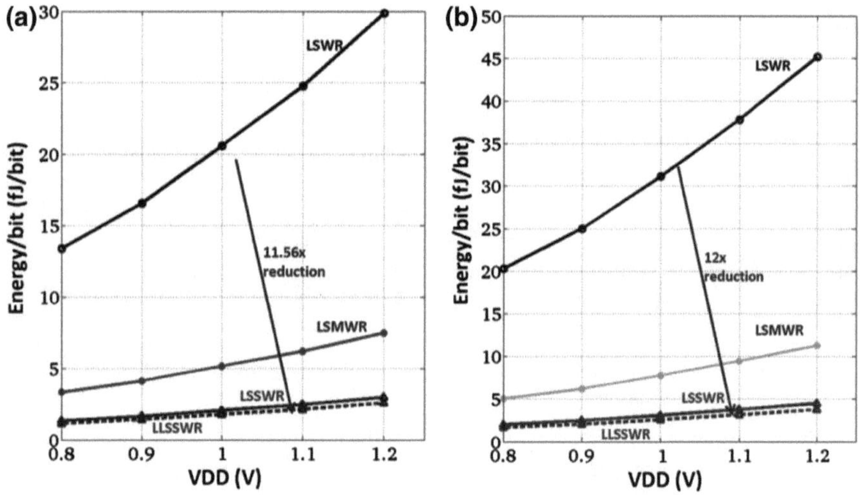

Fig. 5.24 Energy consumption comparison for LSWR. Techniques for the best case scenario [(1/4) un masked quarters for LSMWR, data pattern (10 %, $Prb_{switching}$) for LSSWR]. Column height = 512 cells, nominal process corner, Temp = 25 °C. **a** 40 nm LP. **b** 65 nm LP

capacitances because of the area reduction achieved with the LLSSWR implementation. And the reduced timing complexity associated with the nonstrobed local WR used for the local hierarchy. For the worst case data patterns when only one-fourth of the word length is masked Low Swing WRITE with WRITE Masking (LMSWR) results in 25 % reduction in the energy consumption (Fig. 5.23).

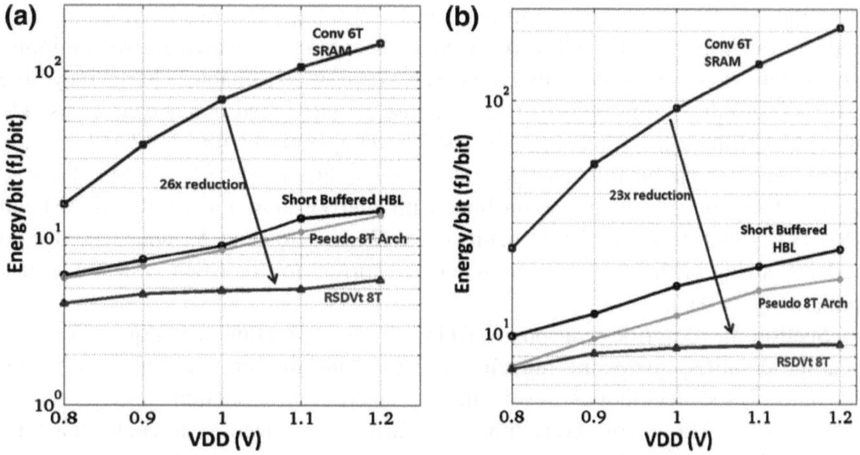

Fig. 5.25 Energy consumption comparison for READ operation. Column height = 512 cells, nominal process corner, Temp = 25 °C. **a** 40 nm LP. **b** 65 nm LP

But LLSSWR achieves (2.1–2.3)x reduction in the energy consumption for 40 nm and 65 nm LP technology. For the best case data patterns when three-fourth of the word length is masked, the energy reduction gain for LSMWR operation is 75 %. Assuming that the probability of switching to be 10 %, then the energy reduction gain with LLSSWR operation can be (11.6–12)x (Fig. 5.24) for 40 nm and 65 nm LP technology.

Figure 5.25 shows READ energy consumption for 40 nm and 65 nm LP technology. RSDVt 8T SRAM results in the most energy efficient READ operation. Reduced swing read bit-line reduces the energy consumption associated with the read bit-lines. Furthermore, the absence of the local assist circuitry with RSDVt 8T rules out the requirement for the complex timing and also helps in reducing the energy consumption. It results in (26–23)x reduction in the READ energy consumption compared to the conventional SRAM 6T cell. Energy reduction gain with pseudo 8T SRAM cell compared to the low swing short buffered SRAM decreases with the technology scaling. This is because of the local sense amplifier used in the local hierarchy. The local sense amplifier has to use upsized transistors because of the increased mismatch offset with the technology scaling Coseman et al. (2009), which reduces the energy reduction benefit.

5.5 Summary

This chapter outlines various circuit techniques for minimizing static and dynamic energy reduction. Section 5.1 describes various leakage reduction techniques. The HVT transistor based SRAM array with gated read buffer reduces leakage by 2.8x in 40 nm LP technology node. The local assist circuitry includes a local sense amplifier on the short LBLs and a gated read buffer. The local sense amplifier reduces the impact of the cell read current on access speed, which allows minimum sized high Vt cell transistors, this reduces leakage. It also enables charge recycling with VDD/2 precharged short LBLs. The use of gated read buffer enables pseudo 8T SRAM (Sharma et al. 2010) cell type READ operation with 6T SRAM cell and also eliminates the bit-line leakage under idle conditions. For 8T SRAM cell based arrays. The reduced swing read bit-lines with dual Vt transistors for the SRAM cell (RSDVt) (Sharma et al. 2011) achieve 7x (2x due to dual Vt and 3.5x due to reduced swing bit-lines) reduction in leakage current for the worst case stored data pattern (Q = 'H').

Section 5.2 describes various WRITE energy reduction techniques. Voltage scaling definitely helps in reducing energy consumption, but the low swing bit-lines based techniques result in the maximum reduction in the energy consumption. The energy consumption is further optimized by having a masking feature. For every masked quarter the energy reduction gain can be between 25 % and 75 % compared to the low swing bit-lines based techniques. Litho optimized low swing static write signal based techniques reduce the timing complexity and transistor sizes. The energy reduction gain is maximum for LLSSWR (Sharma et al. 2012) compared to the other state-of-the-art techniques.

Section 5.3 discusses the local assist circuitry from the energy consumption prespective. Pseudo 8T (Sharma et al. 2010) solves the limitation associated with the conventional local assist circuitry in achieving low energy consumption. It enables low swing signals not only on the global bit-lines but also on the LBLs. The VDD/2 precharged LBLs result in charge recycling and reduces the energy consumption. RSDVt 8T SRAM (Sharma et al. 2011) results in 26x reduction in the energy consumption compared to the conventional 6T SRAM cell in 40 nm LP technology node.

References

K. Agawa, H. Hara, T. Takayanagi, and T. Kuroda, A bit-line leakage compensation scheme for low-voltage SRAMs. IEEE J. Solid-State Circuits. 726–734 (2001)

U. Bhattacharya et al., 45 nm SRAM technology development and technology lead vehicle. Intel Technol. J. **12**(2) (2008)

S. Cosemans, W. Dehaene, F. Catthoor, A low-power embedded SRAM for wireless applications. IEEE J. Solid-State Circuits **42**(7), 1607–1617 (2007)

S. Cosemans, W. Dehaene, F. Catthoor. A 3.6 pJ/Access 480 MHz, 128 kbit On-Chip SRAM with 850 MHz Boost Mode in 90 nm CMOS with Tunable Sense Amplifiers. IEEE J. Solid-State Circuits **44**(7), 2065–2077 (2009)

K. Kanda et al., 90 % Write power-saving SRAM using sense-amplifying memory cell. IEEE J. Solid-State Circuits **39**(6), 927–933 (2004)

H. Kawaguchi et al., Dynamic leakage cut-off scheme for low-voltage SRAMs, Digest of technical papers of the 1998 *Symposium on VLSI Technology*, pp. 140–141, June 1998

K. Kim, H. Mahmoodi, K. Roy, A low-power SRAM using bit-line charge recycling. IEEE J. Solid-State Circuits **43**(2), 446–459 (2008)

T.H. Kim et al., A Voltage Scalable 0.26V, 64kb 8T SRAM With Vmin Lowering Techniques and Deep Sleep Mode. IEEE J. Solid-State Circuits **44**(6), 1785–1795 (2009)

Y.C. Lai and S.Y. Huang, X-Calibration: a technique for combating excessive bit-line leakage current in nanometer SRAM Designs. IEEE J. Solid-State Circuits. 1964–1971 (2008)

K. Mai et al., Low-power SRAM design using half-swing pulse-mode techniques. IEEE J. Solid-State Circuits **33**(11), 1659–1671 (1998)

V. Sharma et al., A 4.4 pJ/access 80 MHz, 2 K word X 64 b memory with write masking feature and variability resilient multi-sized sense amplifier redundancy for W.S. nodes. Proc. ESSCIRC, pp. 358–361, Sept 2010

V. Sharma et al., 8T SRAM with mimicked negative bit-lines and charge limited sequential sense amplifier for wireless sensor nodes. in *Proceedings of IEEE European Solid State Circuits Conference (ESSCIRC)*, pp. 531–534, Sept 2011

V. Sharma et al., *Ultra Low Power Litho Friendly Local Assist Circuitry For Variability Resilient 8T SRAM Cell, Design Automation and Test in Europe (DATE)* (Dresden, March, 2012), pp. 11–17

M. Sinangil, N. Verma and A. Chandrakasan, Reconfigurable 8T ultra dynamic voltage scalable (U-DVS) SRAM in 65 nm CMOS. IEEE J. Solid-State Circuits. 3163–3173 (2009)

N. Verma, A. Chandrakasan, A 256kb 65nm 8T Subthreshold SRAM Employing Sense-Amplifier Redundancy. IEEE J. Solid-State Circuits **43**(1), 141–149 (2009)

M. Yamaoka et al., 90-nm process-variation adaptive embedded SRAM modules with power-738 line-floating write technique. IEEE J. Solid State Circ. **41**(3), 705–711 (2006)

B.D. Yang, L.S. Kim, A low-power SRAM using hierarchical bit line and local sense amplifiers. IEEE J. Solid-State Circuits **40**(6), 1366–1376 (2005)

K.H. Kim et al., A Voltage Scalable 0.26V, 64kb 8T SRAM With Vmin Lowering Techniques and Deep Sleep Mode. IEEE J. Solid-State Circuits 44(6), 1785–1795 (2009)

[illegible]

Chapter 6
Variation Tolerant Low Power Sense Amplifiers

This chapter describes the READ sense amplifier (SA) of the memory. It discusses the fundamental limitation on the SA performance, especially for the memories in deep sub micron technologies. It covers various calibration based SA design techniques. With the practical implementation details for Multi-sized SA redundancy. And comparison of the various calibration based techniques. Then a charge limited sequential sensing concept is discussed. Finally the design and implementation details of a calibration free SA based on the charge limited sequential sensing is provided.

6.1 Introduction: Energy-Offset Trade off Problem in Sense Amplifier Circuits

A SA resolves a small input voltage difference applied to its input terminals to a full swing voltage level output. The READ SA is designed to sense the low voltage swing created by the small sized SRAM cell on the bit-lines. The reduction in low voltage swing created by an accessed SRAM cell reduces the energy consumption associated with charging and discharging of high capacitive bit-lines (Fig. 6.1). Technology scaling results in decreasing cell read current and the bit-line capacitance is not scaling (reducing) proportionally. This results in access speed degradation for a given number of SRAM cells. With low swing bit-lines, the accessed SRAM cell has to develop less swing thereby decreasing the memory access time. Therefore, small bit-line swing sensing (<100 mV) is more desirable for the advanced technology nodes. But the minimal bit-line swing that can be resolved reliably is limited by the offset of the SA.

The increased random Vt variations with technology scaling also increases the mismatch offset voltage. The increased mismatch offset puts a limit on the minimum input bit-line swing that can be sensed reliably. The correct functionality

V. Sharma et al., *SRAM Design for Wireless Sensor Networks*, Analog Circuits and Signal Processing, DOI: 10.1007/978-1-4614-4039-0_6,

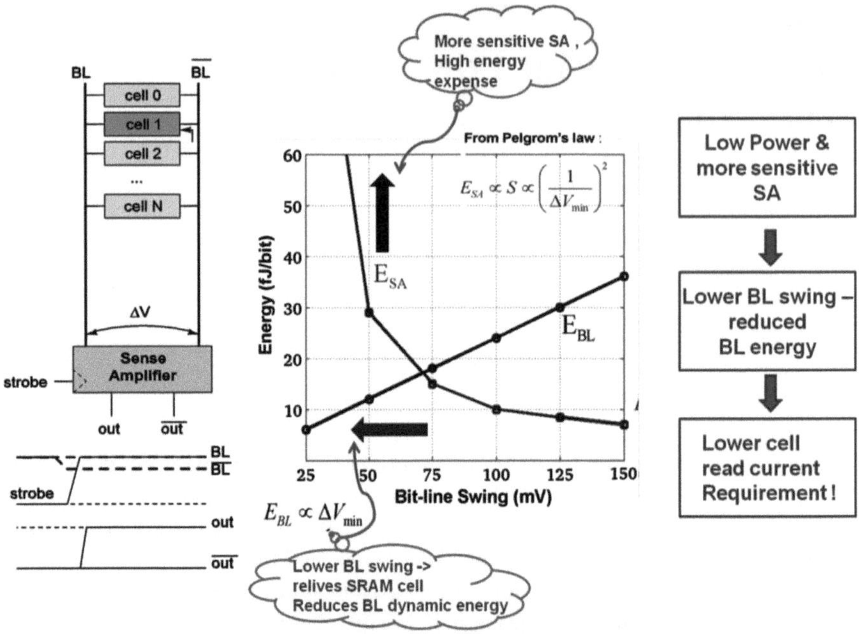

Fig. 6.1 Energy versus sensitivity trade-off for SAs. A lower offset voltage allows the use of a lower BL swing, but to achieve that, the SA energy increases quadratically

of the SA circuit is ensured by upsizing the critical transistors (reduces the mismatch offset). The size of the critical input transistors of SA increases quadratically with the input bit-line swing. The traditional method for reducing the mismatch offset relies on upsizing the critical transistors. The increased transistor sizes in order to enable a low swing bit-line sensing directly increases the energy consumption of a SA circuit. Therefore, SAs are becoming critical feature in SRAM design for achieving ultra low energy operation for the advanced technology nodes.

The minimal target value of the required bit-line discharge ($\Delta V_{\min}$) depends on the technology, SA design, sizing, and the target yield level. If V_{offset} follows a Gaussian distribution with 0 mean and a standard deviation σ_{offset}, then the required minimum bit-line discharge computed from the inverse of the normal cdf. The desired yield target is expressed in the number of standard deviations, n. For the above example $n = 5.33$.

$$\Delta V_{\min} = F^{-1}\left(f_r/2 | \mu, \sigma\right) = 5.33\sigma \tag{6.1}$$

According to the Pelgrom law (Pelgrom et al. 1989) the minimum bit-line discharge for a given reliability requirement which can be resolved by the input transistors of a SA is.

$$\Delta V_{\min} \geq n_{\mathrm{Fr}} \times \sigma_{\mathrm{offset}} \approx \frac{n_{\mathrm{Fr}} \times A_{\Delta\mathrm{Vt}}}{\sqrt{W \times L}} \tag{6.2}$$

n_{Fr} the desired yield (reliability requirement) expressed in the number of standard deviations

σ_{offset} the standard deviation of the difference in the threshold voltage ΔVt between the 2 input transistors

$A_{\Delta\mathrm{Vt}}$ Pelgrom constant

W width of the SA input transistor

L length of the SA input transistor

$\sigma_{\mathrm{V_T}}$ the standard deviation of the difference in the threshold voltage between the 2 minimal transistors is.

$$\sigma_{V_T} = \frac{A_{\Delta\mathrm{Vt}}}{\sqrt{W_{\min} \times L_{\min}}} \tag{6.3}$$

The upscale factor S for the input transistor pair is defined as

$$S = \frac{W \times L}{W_{\min} \times L_{\min}} \tag{6.4}$$

Reformulating Eq. 6.2 based on Eqs. 6.3 and 6.4.

$$\sqrt{S} \geq \left(\frac{n_{\mathrm{Fr}} \times \sigma_{V_T}}{\Delta V_{\min}}\right) \tag{6.5}$$

The size of a strobed SA is quadratically proportional with the reliability requirements for a fixed input swing. The energy consumption is directly proportional to the sizing requirements. In other words, the reliability margins directly impacts the energy consumption of a SA. For e.g., designing for $n_{\mathrm{Fr}} = 6.1 (f_r = 10\mathrm{e}-9)$ will increase the SA energy by 155 % compared to the one designed for $n_{\mathrm{Fr}} = 4.9$ $(f_r = 10\mathrm{e}-6)$.

In traditional SA design this offset is reduced by increasing the size of the critical transistors (Pelgrom et al. 1989), which directly maps into an increased dynamic energy consumption. This is becoming problematic, especially for memories with large word length designed in deep submicron technologies. The expected contribution of SA energy to the total READ energy of the memory is expected to increase to 29 % in 32 nm (Cosemans et al. 2009) for resolving 100 mV compared to 90 nm technology node.

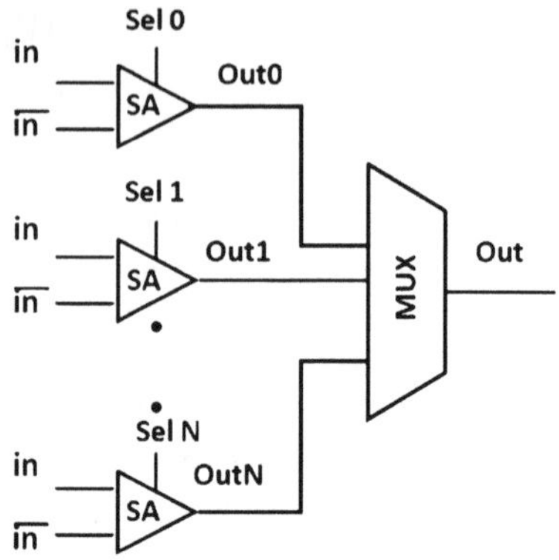

Fig. 6.2 SA redundancy (Verma and Chandrakasan 2008)

6.2 Calibration Based Techniques.

SA calibration is a family of techniques that solves the SA mismatch offset problem enabling a low input swing sensing with minimal impact on the energy consumption and the sensing delay. This section conceptualizes various calibration techniques with the implementation details for Multi-Sized SA redundancy (MS-SA-R) (Sharma et al. 2010). Cosemans (2009) provides a detailed quantitative analysis for the calibration techniques and is not repeated in this section.

6.2.1 Sense Amplifier Redundancy

In traditional memory design there is a single SA per bit-line designed for a certain failure rate f_r. In Verma and Chandrakasan (2008) this single SA is replaced by set of N equal sized smaller SAs. Figure 6.2 shows SA redundancy. There is a separate calibration phase to find a working SA from the set. Then only this selected SA is activated during the READ operation. Under the assumption that SA failures are independent, with N-fold redundancy, the SA size can be relaxed to S $(\{f_r\}^{1/N}, \Delta V_{\min})$ compared to S $(\{f_r\}, \Delta V_{\min})$ without calibration. The reduced SA size maps into reduced energy consumption.

6.2.2 Sense Amplifier Tuning

In Cosemans et al. (2009) voltage tuning is used for the offset cancellation. Each SA receives the most appropriate reference voltage based on the offset. Figure 6.3 shows SA tuning. For a given SA design, the minimal required input signal is N times smaller compared to the one with no tuning. N refers to the number of reference voltage levels available. SA tuning does not require any selection in the

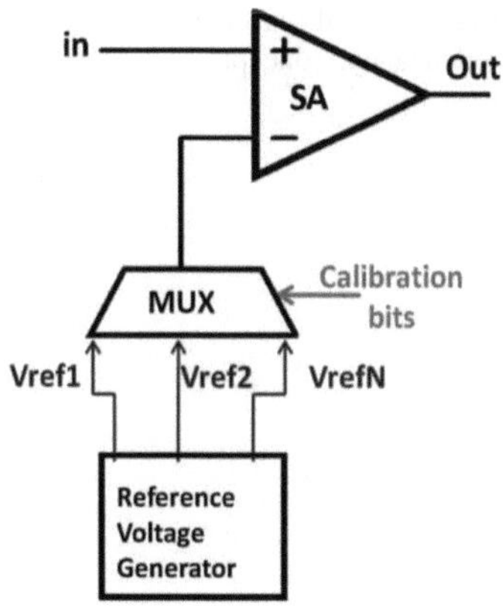

Fig. 6.3 Sense amplifier tuning (Cosemans et al. 2009)

critical logic path, so it scales gracefully to large values of *N* compared to SA redundancy. *N* refers to number of reference voltages for the tuning and number of redundant SAs for SA redundancy. SA tuning definitely reduces the active energy consumption, but it does not take into account the energy consumption required to generate multiple reference voltages and also the increase in area with this approach. Alternatively, Sinangil et al. (2009) limits the reference voltage choice to only 2 values.

6.2.3 Capacitive Resist Implementation and Parallel Device Assist Implementation

Bhargava et al. (2009) proposes capacitive resist implementation (CRI) and parallel device assist implementation (PDAI) for offset compensation. In CRI, switched capacitors are used to decelerate the fast charging and discharging side (under the impact of Vt mismatches) of the regenerative feedback cross-coupled. The capacitor slows down the faster side by increasing the capacitance that needs to be discharged. The switches are implemented using PMOS devices and the capacitors are source drain shorted PMOS devices. The tuning is controlled by varying the size of the capacitor and varying the size of the switch.

Alternatively, in PDA1 a pair of NMOS devices are placed in parallel with the regular pair of pull down NMOS. By turning on the NMOS parallel to the slower pull down aids in assisting the weaker side in faster resolution and hence compensate the mismatch offset. With PDAI tuning is controlled by varying the size of the parallel conducting device. It proposes to use 2 bits for configuration: no kick, positive kick, negative kick, and double kick.

Fig. 6.4 Multi-sized SA redundancy: concept

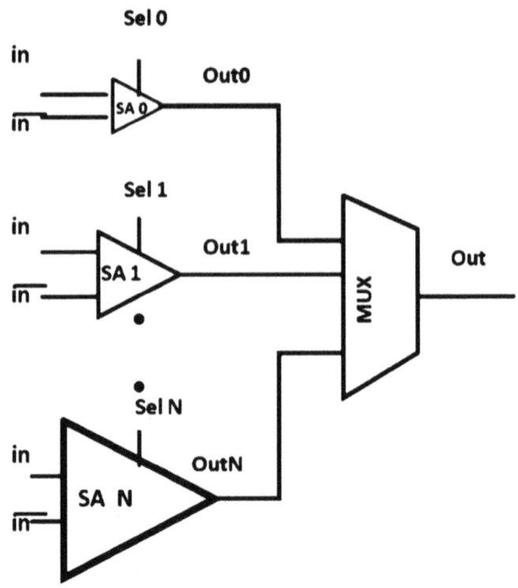

6.2.4 Hot Carrier Injection Trimming

Kawasumi et al. (2010) proposes hot carrier injection (HCI) for mismatch reduction. However, this technique requires very high supply voltage levels (3.0 V) and therefore is not suitable for low energy applications. Also the Vt shift on the cross-coupled PMOS transistors caused by HCI or negative bias temperature instability (NBTI) is an issue with this technique.

6.2.5 Multi-Sized SA Redundancy

(1) **Concept and Implementation**

MS-SA-R replaces a single SA with a set of different sized SAs, collectively having the same or even less f_r. In other words, N redundant differently sized SAs are used, with failure rates $fr_1 \ldots fr_N$. The indices are sorted on SA sizes so that the smallest SA size with maximum failure rate has index 1. Figure 6.4 shows concept of MS-SA-R.

For example, a traditional SA designed for 6σ yield is replaced by two SAs. The smaller one is designed for 2σ yield, with its energy consumption 9 times smaller than that of the traditional single SA (E_{trad}) designed for 6σ yield. The bigger SA in a set is designed for 6σ yield. Then the average energy consumption (E_{Avg}) of MS-SA-R, $N = 2$ is approximately 6 times smaller compared to that of the traditional SA. The sizes of redundant multi-sized SAs are based on the metric of minimizing the total energy consumption of the SA system for a target yield requirement. When the optimal sizes of the critical transistors of the smallest SA approach closer to the minimum transistor sizes of a technology, it determines an upper limit on the value of N for MS-SA-R. Also the area overhead of selection

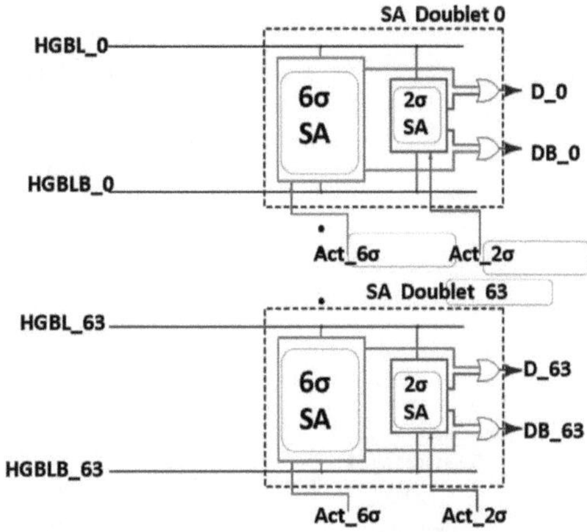

Fig. 6.5 Implementation of multi-sized SA redundancy

logic with the larger values of N cannot be ignored. MS-SA-R in (Sharma et al. 2010) is implemented with $N = 2$. The doublet of SA used consists of a smaller SA designed for 2σ yield and the other SA designed for 6σ yield. There is a separate calibration phase to find a working SA from the set. Then, only this selected SA is activated during the READ operation. Figure 6.5 shows MS-SA doublet with calibration bits.

(2) **Calibration Algorithm used with Multi-Sized SA Redundancy**

The calibration algorithm selects a SA out of the redundant set of differently sized SAs. Figure 6.6 shows calibration algorithm for MS-SA-R . It is desired to use each time the smallest available SA which does not fail. The calibration is done as follows

Step (1) Initially only the 2σ sized SA are selected {Act_2σ high} and the predefined data stored is successfully read.
Step (2) A READ operation with 2σ SA is performed. Failing bit locations are identified. For example, when reading a 64 bit word with bit locations 2, 5, 31, and 64 are erroneous, determined by comparing with the predefined data stored, then for the doublet numbers 2, 5, 31, and 64 the 6σ SA is selected.
Step (3) The calibration register is updated with the selection of 6σ SA information. Then READ operation is performed with 2σ SA enabled except for the number 2, 5, 31, and 64 doublets for which Act_6σ is enabled.
Step (4) If step 3 confirms step 2 (successful READ operation), then the selection bits will be locked in the calibration, till the next calibration phase. In the example, with the 2σ SAs of doublet numbers 2, 5, 31, and 64 failing, this means that the following selection is stored:

{0110111111111111111111111111110111111111111111111111111111111110}

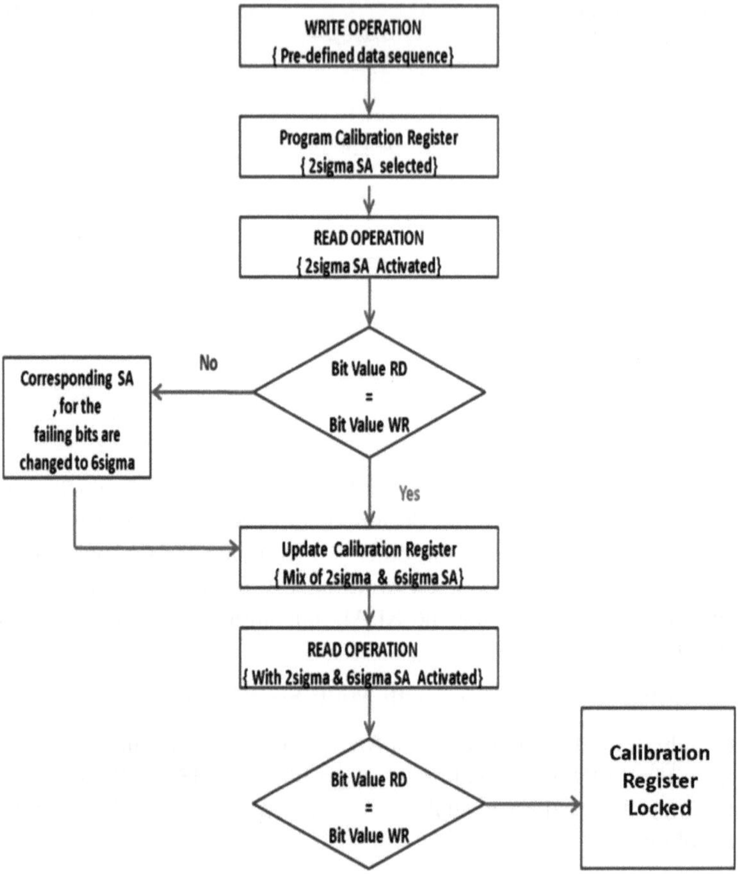

Fig. 6.6 Calibration algorithm for multi-sized SA redundancy

wherein 0 represents Act_6σ high and 1 represents Act_2σ high. These values will be used for the memory accesses, till the next calibration phase. The next calibration phase is determined based on the user application.

6.3 Charge Limited Sequential Sense Amplifier: Calibration Free Solution

This section explains the proposed novel charge limited sequential SA design (CLS-SA). Section 6.3.1 discuss the limitations associated with the low input swing, low energy calibration based SA design for the advanced technology nodes. Section 6.3.2 explains the concept of charge limited sequential sensing. Section 6.3.3 discusses the circuit-level design of SA based on this concept and Sect. 6.3.4 illustrates the basic operation of CLS-SA

6.3.1 Limitations with the Calibration Based SA Design

Calibration techniques are quite effective, but for many applications, the introduction of a separate calibration phase is not acceptable. Every test vector applied to calibrate the SA increases the test cost and test time. This inhibits the advantages offered by the calibration techniques and is not an option for low energy SoC designs. Therefore, an alternative calibration free technique to improve the energy-offset trade-off is required. In this work, a novel SA CLS-SA is proposed, which resolves the energy-offset tradeoff issue without resorting to the post-silicon tuning.

6.3.2 Charge Limited Sequential Sensing: Concept

In normal SA designs, the critical transistors are made large to reduce the offset voltage. This makes the SA sensitive, but this also results in large capacitances on the internal nodes. In traditional designs, each sensing decision requires that one of those nodes makes a full swing transition, which results in a large energy consumption. In the CLS-SA, two SA stages are used. The critical transistors of the first stage are about the same size of those of the traditional design. However, the voltage swing on the internal nodes is limited, e.g. to $(4-5)\times$ the initial input signal. After the first stage sensing is completed, the signal on the internal nodes of this first stage is used as input for the second stage. This second stage has a larger input signal and can hence be much smaller, with smaller capacitances on the internal nodes. This second SA amplifies the signal to a full logic level. The energy consumption expense due to the large capacitances on the internal nodes of first SA is reduced by limiting its output swing. This first SA acts as a pre-amplifier with a limited output swing. The limited output swing of the first SA is then resolved sequentially to a full voltage swing by a much smaller second SA. As this second SA is small, it has low energy consumption. The pre-amplification information of the first SA is available on its internal nodes instead of on the bit-lines. The input of the second SA is directly connected to the output of the first SA.

Sizing of a traditional SA. If the size of the SA is expressed as S_{SA} multiples of the minimal transistor size, the minimal required size for a traditional SA design can be calculated from Pelgrom's law:

$$(S_{SA}) \geq \left[\frac{n_{Fr} \times \sigma_{V_T}}{\Delta V_{min}}\right]^2 \tag{6.6}$$

Here, n_{Fr} is the number of standard deviations that is needed as margin to achieve a failure rate below Fr. Assuming a Gaussian distribution, $n_{Fr} = 6.1$ corresponds to about one failure in 10^9 SAs. σ_{VT} is the standard deviation of the difference in threshold voltage between 2 minimal size transistors and V_{min} is the smallest signal the SA must reliably resolve. In a classic design this value is set to the effective input signal that is available for the SA. A traditional SA with complementary read bit-lines compares the voltage on the bit-line with the voltage on the complementary bit-line, hence V_{min} is the bit-line swing. A SA with a single bit-line has to compare the bit-line voltage with a reference voltage, so $V_{\text{min}} = \text{BL}$ swing/2.

Design of the CLS-SA. The required size of the first stage of the CLS-SA ($S_{\text{first\_SA}}$) is about the same as that of the traditional SA:

$$(S_{\text{first\_SA}}) \geq \left(\frac{n_{\text{Fr}} \times \sigma_{V_T}}{\Delta V_{\text{min}}}\right)^2 = \left(\frac{n_{\text{Fr}} \times \sigma_{V_T}}{\Delta V_{\text{in,first}}}\right)^2 \tag{6.7}$$

$V_{\text{in,first}}$ is the input voltage signal for the first SA, which is equal to V_{min}.

The first SA amplifies the signal with a factor A, which results in an output swing $\Delta V_{\text{out,first}}$ on its nodes (not on the bit-lines)

$$\Delta V_{\text{out,first}} = A \times \Delta V_{\text{in,first}} \tag{6.8}$$

This output signal of the first SA is used as the input signal for the second SA

$$\Delta V_{\text{in,second}} = \Delta V_{\text{out,first}} \tag{6.9}$$

The minimal required size of the second SA ($S_{\text{second\_SA}}$) is then

$$(S_{\text{second\_SA}}) \geq \left(\frac{n_{\text{Fr}} \times \sigma_{V_T}}{\Delta V_{\text{in,second}}}\right)^2 = \left(\frac{n_{\text{Fr}} \times \sigma_{V_T}}{A \times \Delta V_{\text{in,first}}}\right)^2 \tag{6.10}$$

The first and second SA are designed for the same target yield ($n_{Fr} = 6$). As the second SA is designed with respect to the larger input swing $\Delta V_{\text{in,second}}$, the size of the second SA is (A^2) times smaller than that of the first SA.

$$S_{\text{second\_SA}} = \frac{S_{\text{first\_SA}}}{A^2} \tag{6.11}$$

The energy consumption of the CLS-SA is hence

$$E_{\text{Total,CLS-SA}} = E_{\text{first\_SA}} + E_{\text{second\_SA}}$$

$$\begin{cases} E_{\text{first\_SA}} & = & C_{\text{first\_SA}} \times \text{Vdd} \times (A \times \Delta V_{\text{in}}) \\ E_{\text{second\_SA}} & = & C_{\text{second\_SA}} \times \text{Vdd} \times (\text{Vdd}) \\ E_{\text{Traditional\_SA}} & = & C_{\text{first\_SA}} \times \text{Vdd} \times (\text{Vdd}) \end{cases}$$

$$E_{\text{Total,CLS-SA}} = \text{Vdd} \times C_{\text{first\_SA}} \times \left(A \times \Delta V_{\text{in}} + \frac{1}{A^2} \times \text{Vdd}\right) \tag{6.12}$$

The overall failure rate of CLS-SA is approximately the sum of the failure rate of the first SA and the failure rate of the second SA. The optimal amplification A for CLS-SA, with corresponding sizes and energy improvement with CLS-SA is shown in Fig. 6.7b. Figure 6.7c shows the energy comparison of a bit-line with

CLS-SA and a bit-line with traditional SA for optimal A. Figure 6.7d shows bit-line discharge time. Figure 6.7e illustrates trade-off between energy and delay for CLS-SA and traditional SA. For a bit-line swing of 106 mV bit-line energy consumption with CLS-SA is 1.65x lower compared to the traditional SA with 13.3 % increase in delay.

6.3.3 Circuit Implementation

Basic Big_SA topology. The SA design based on the concept of charge limited sequential sensing is shown in Fig. 6.8. The first SA (Big_SA) consists of only the PMOS cross-coupled pair rather than the entire structure of the voltage latch type SA. The offset requirement for the second SA (small_SA) is determined by the voltage difference between the two output nodes of the Big_SA rather than by the absolute potential on the output nodes. The NMOS transistors can be omitted from Big_SA as long as the gain achieved by the PMOS differential pair is sufficiently large and the required output signal ($\Delta V_{\text{out,first}} = \text{A} \times \Delta V_{\text{in,first}}$) is sufficiently small compared to Vdd. Omitting the NMOS transistors reduces the energy and area consumption, and avoids the additional offset that would be contributed by the Vt mismatch of the NMOS transistors.

Impact of variability. When the output swing of Big_SA is controlled by limiting the activation time of the current source PMOS (Mtop) which feeds Big_SA, intra-die, inter-die, and temperature variations result in a huge spread of the output swing of the Big_SA. The first SA must be activated during a sufficient time period to ensure that the resulting output swing is sufficiently large under all conditions. The time required to generate the output signal depends mainly on the Ion of the current source. This minimal required time period is defined by the worst-case combination of process corner and transistor mismatch. However, during this time period, all other SAs will develop a much larger output voltage swing. This would severely impact the efficiency of limited swing sequential sensing.

Charge limited voltage swing. This problem is remedied by replacing the power rail of the Big_SA with a capacitor (C_{source}) (Fig. 6.8), which is pre-charged before the sensing operation starts.

Optimizing Csource. The value of the capacitor is chosen just large enough to ensure sufficient output swing for the worst-case SA. Figure 6.9a, shows the latency between activation of Big_SA and the time at which its output swing has reached the value $4 \times V_{\text{in}}$ as function of Cs. Increasing C_s improves the sensing speed but increases the energy consumption (Fig. 6.9b). For the WSN nodes design, the energy constraint is more stringent than the latency constraint. A small value, ~8 fF for source capacitance is recommended. The first stage of the CLS-SA is activated by enabling Mtop, after disconnecting the internal nodes from the bit-lines. This initiates the sensing operation. Mtop provides charge from

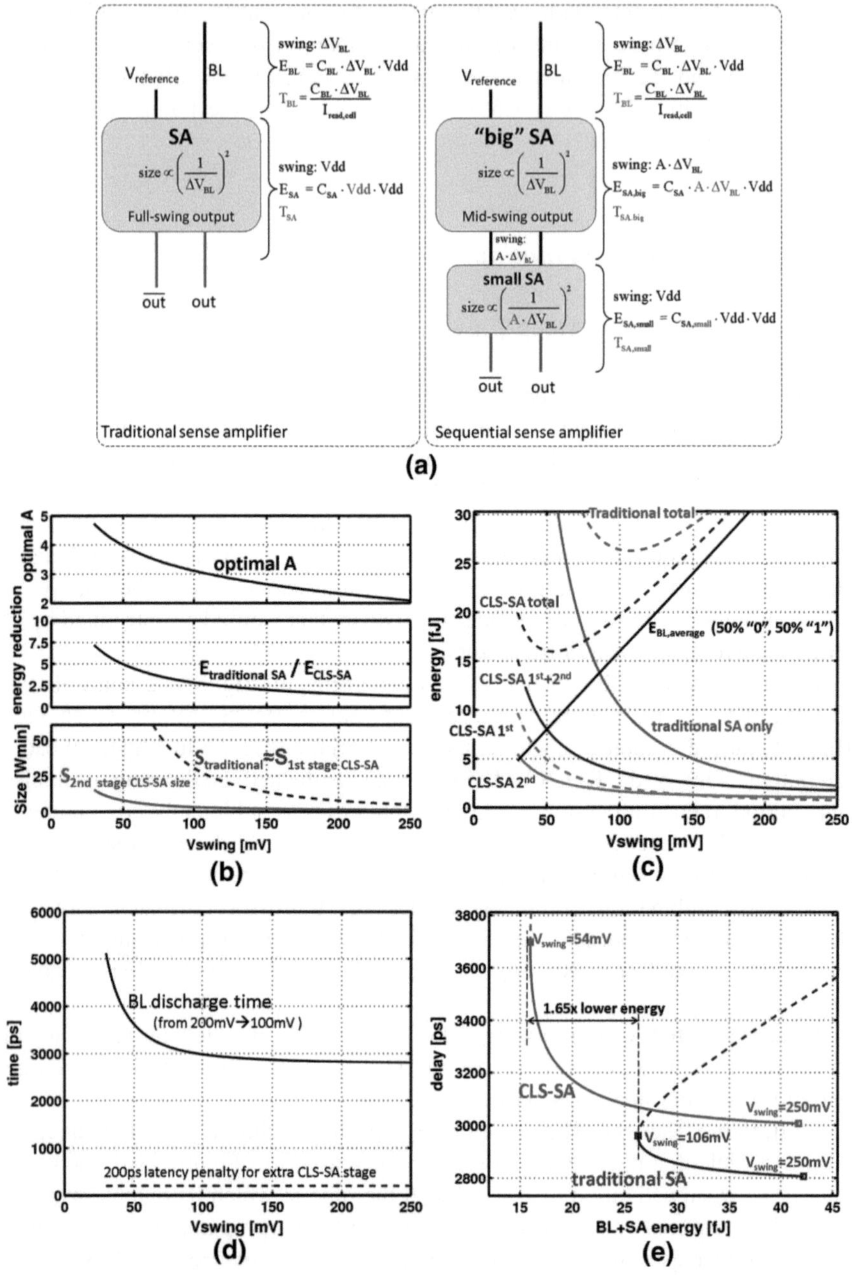

Fig. 6.7 Optimization and performance of charge limited sequential sense amplifier (CLS-SA) according to Eqs. 6.1–6.7 for a setup with Vdd = 0.8 V, $n_{fr} = 6$, $R_{\text{cell}} = 20\ \text{k}\Omega$, $C_{\text{BL}} = 200$ fF, single-ended bit-line with $V_{\text{BL pre-charge}} = V_{\text{BL,discharge}} = 2 \times V_{\text{swing}}$ and 50 % "1" values. **a**. Concept. **b**. Optimal amplification A, corresponding sizes and energy improvement. **c**. Energy for optimal A. **d**. Bit-line discharge time. **e**. Possible trade-off between energy and delay

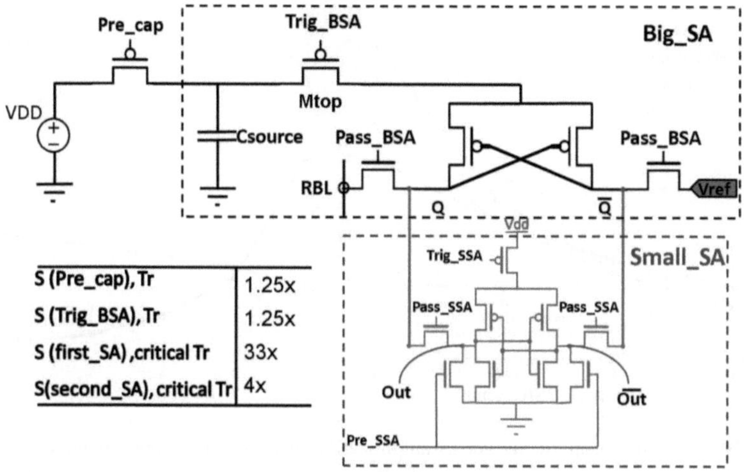

Fig. 6.8 Circuit implementation of CLS-SA (Sharma et al. 2011)

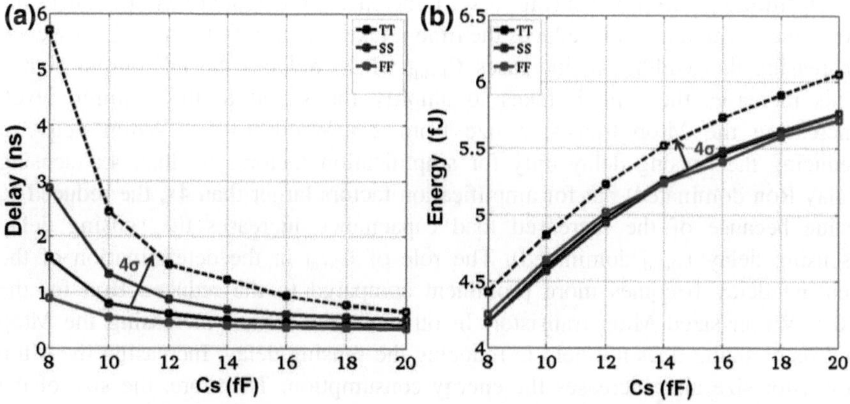

Fig. 9 **a**. Mean sensing delay. **b**. Mean sensing energy consumption of preamplifier circuit (Big_SA) for $V_{in} = 50$ mV and $A = 4$x for different values of source capacitors

C_{source} to the cross-coupled PMOS pair, which uses this charge to amplify the signal on the internal nodes.

After a sufficient time, the final voltage V_f on the signal nodes of Big_SA can be approximated by considering a charge sharing operation between the source capacitance C_{source} and the load capacitance C_{load}, the overall capacitance of Big_SA. This charge sharing results in

$$V_f = \frac{C_s \times \text{VDD}}{(C_s + C_{load})} \tag{6.13}$$

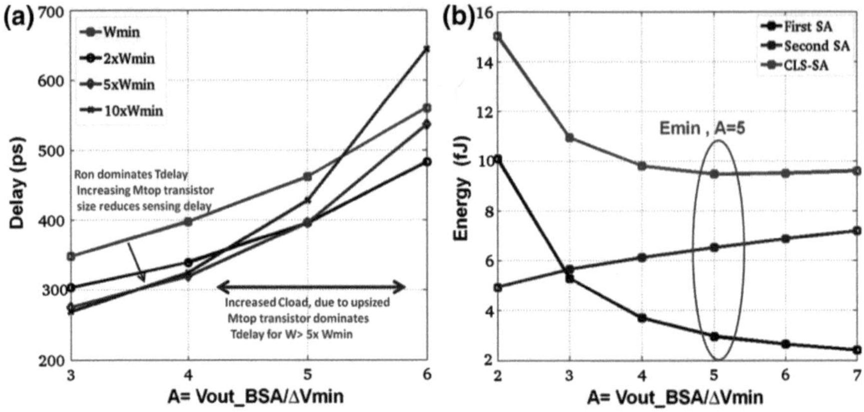

Fig. 6.10 **a**. Mean sensing delay versus amplification factor (A) for input voltage difference of 50 mV for different sizes of Mtop transistors. **b**. Energy consumption versus amplification factor (simulation). Minimum energy consumption for optimum value of $A = 5X$

Optimizing Mtop width (Fig. 6.10a). Increasing the width of Mtop reduces the resistance, which in itself reduces the time it takes to amplify the signal. However, increasing this width also increases C_{load}, which reduces the DC output signal. This increases the time it takes to amplify the signal to the required level. Increasing the Mtop transistor size from 2 × Wmin to 5 × Wmin helps in reducing the sensing delay only for amplification factors less than 4x (sensing delay Ron dominated) but for amplification factors larger than 4x, the reduced V_f value because of the increased load capacitance increases the sensing delay (sensing delay C_{load} dominated). The role of C_{load} in the determination of the sensing delay becomes more prominent compared to the reduced Ron for the 10 × Wmin sized Mtop transistor. In other words, further increasing the Mtop transistor sizing does not help in reducing the sensing delay. Increasing the Mtop transistor size also increases the energy consumption. Therefore, the size of the Trig_BSA activated Mtop transistor is kept minimum (1.25 × Wmin) in this implementation in order to reduce the energy consumption for the given target timing budget.

Optimizing amplification factor A (Fig. 6.10b). The cross-coupled PMOS pair of Big_SA is designed for $n_{Fr} = 6$ with respect to the input voltage difference V_{in}. The critical transistors of the second SA (Small_SA) are designed for the same target yield $n_{Fr} = 6$, but with respect to the input swing $A \times V_{\text{in}}$. Figure 6.13b shows the optimal amplification factor for an input signal $\Delta V_{\text{in}} = 50$ mV. The optimal energy point is reached at $A = 5$x.

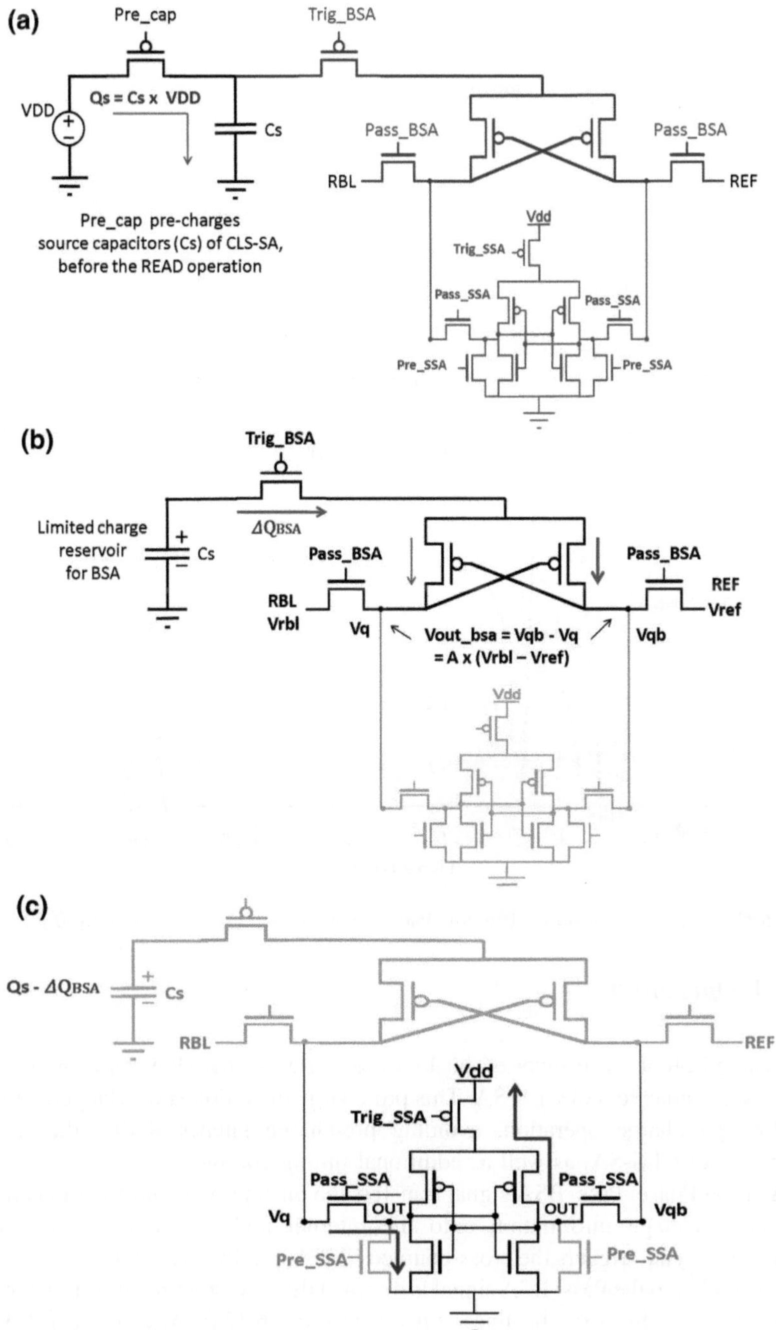

Fig. 6.11 CLS-SA Operation

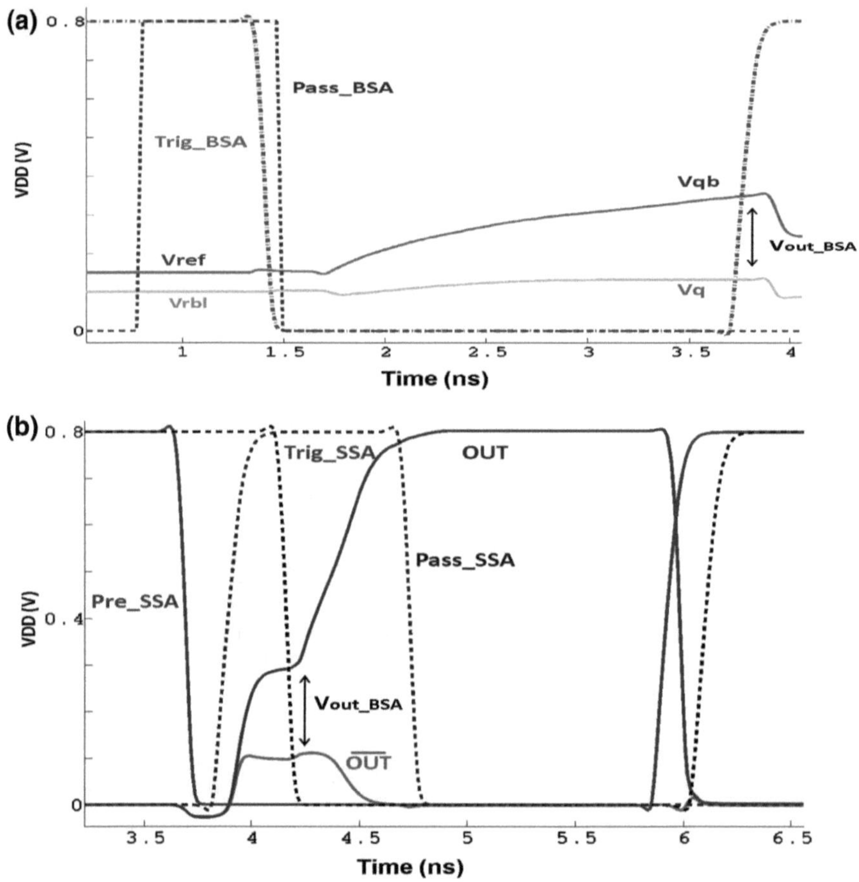

Fig. 6.12 Timing waveform for Big_SA. Partial amplification (V_{out_BSA}) done by Big SA

6.3.4 Operation

Pre-charge phase: The inverted bit-line pre-charge signal (Pre_cap), pre-charges the source capacitors of CLS-SA. This pre-charge operation is overlapped with the bit-line pre-charge operation, avoiding pre-charge latency during the sensing operation of CLS-SA, as well as additional timing complexity.

Sensing Phase: Pass_BSA signal transfers the bit-line swing information and the reference voltage information onto the internal nodes of the Big_SA. Then Trig_BSA signal triggers the cross-coupled PMOS pair by connecting to the charge reservoir Cs and the Pass_BSA signal is disabled thereby isolating the internal nodes of the Big_SA from the bit-lines (Figs. 6.11b and 6.12a). After the Big SA has resolved the low swing input V_{in} to A × Vin, the Trig_BSA signal is disabled. The amount of charge Q_{BSA} used during pre-amplification is restored back on the source capacitors during the next pre-charge phase. The pre-amplified signal on the output

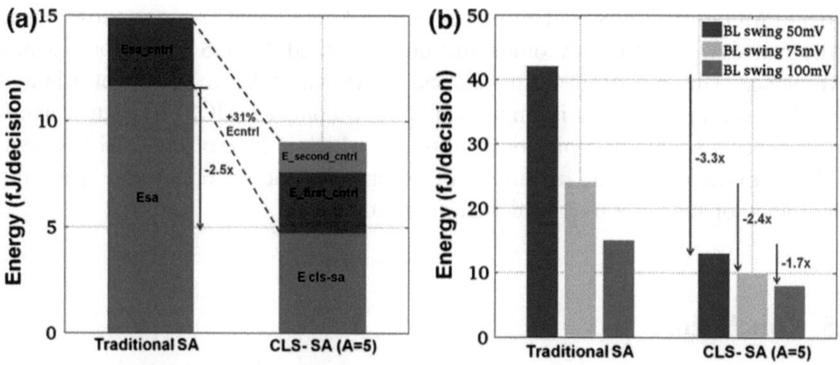

Fig. 6.13 **a**. Distribution of energy consumption of traditional SA and CLS-SA for the bit-line swing of 100 mV. CLS-SA results in 2.5x reduction in the energy consumption. But the additional control signals required for CLS-SA results in 31 % increase in the control energy consumption. **b**. Energy comparison of CLS-SA with traditional SA, including the control signals energy consumption for different values of bit-line swing

nodes of Big_SA is then transferred to the small_SA of Big_SA is then transferred to the Small_SA (Figs. 6.11c and 6.12b). The pre-charge signal of the second SA (Pre_SSA) used for initializing the output nodes is disabled and the Pass_SSA signal transfers the output information of Big_SA onto the internal nodes of the Small_SA. Then the Small_SA is triggered by Trig_SSA, it amplifies the signal (V_{out_BSA}) to a full swing voltage level.

6.4 Comparison

This section discusses only performance CLS-SA. The performance comparison of various calibration based SA design techniques is provided in (Chap. 5 of Cosemans 2009).

If the energy associated with the control signals is neglected, the optimized variability-resilient CLS-SA achieves a 2.5x reduction in the energy consumption compared to the traditional SA design for the same target yield (Fig. 6.13a). The additional control signals required for CLS-SA results in 31 % increase in the control energy consumption. The energy consumption associated with the control signals of Small_SA (transistors sized for $A \times V_{in}$) is 3× less than the energy consumption associated with the control signals of the Big_SA (transistors for V_{in}). Figure 6.13b shows the energy reduction achieved with CLS-SA for different bit-line swings. The energy reduction (including the control overhead) compared to a traditional SA design is larger as the bit-line swing becomes smaller: e.g. 1.7× for a 100 mV BL swing and 3.3× for a 50 mV BL swing. Low bit-line swings are desirable because they reduce the cell latency. For designs that target small input signals, the CLS-SA area is comparable to that of traditional SA designs. The critical PMOS transistors of the first SA (Big_SA) are about as big as those of the traditional SA, but Big_SA omits the cross-coupled NMOS transistors, which saves space. The transistors for the second SA (Small_SA) are almost of minimal size. The source capacitor is implemented as a compact MOSFET gate capacitance; hence Csource occupies only 1.46 um^2 which is 5 % of the CLS-SA area. With the extended bit-line hierarchy, in which only one set of SAs are used, the importance of this area overhead is further reduced.

6.5 Conclusion

This chapter discusses various calibration based techniques for solving energy–offset tradeoff for SA design. Redundancy replaces a traditional SA with multiple SAs of the same size (Verma and Chandrakasan 2008) or with multiple SAs with individually optimized sizes to achieve minimal energy consumption (Sharma et al. 2010). There is a separate calibration phase, during which the system determines which SA of the set to be used. SA-Tuning (Cosemans et al. 2009; Sinangil et al. 2009) provides different values to be used as reference voltage for the SA to compensate for the offset of the SA. Bhargava et al. (2009) proposes CRI and PDAI for mismatch offset compensation. There is a separate configuration register enable possible settings for the mismatch offset compensation. Kawasumi et al. (2010) uses selective HCI to counteract the intrinsic mismatch of the critical SA transistors after fabrication. The transistor threshold voltage shift caused by the HCI is used for mismatch offset reduction.

MS-SA-R as implemented in (Sharma et al. 2010) achieves the maximum reduction in the energy consumption of the global read SAs as compared to the other calibration based techniques for the same number of calibration bits. There is no requirement of high on chip voltages and the risk of Vt shift on the cross-coupled PMOSs owing to HCI or NBTI, the case with HCI trimming (Kawasumi et al. 2010). But for certain applications having a dedicated calibration phase are not desirable because of an increased test cost and test time.

CLS-SA presents an alternative technique to improve the energy-offset trade-off. The CLS-SA uses two SA stages. The first, large SA senses the bit-line signal and amplifies it with a limited amplification factor, for example 4 or 5. This amplified signal is then provided as input to the second SA, which can be much smaller because of its larger input signal. The second SA amplifies the signal to full logic levels. The critical transistors of the first stage must be sized as those of a traditional SA design. However, as the voltage swing on the internal nodes is limited, its energy is much lower. As the second stage has a larger input signal, it is much smaller than the traditional SA. The total energy consumption of the CLS-SA is significantly lower than that of the traditional implementation. At low voltages, process and temperature variations make it difficult to accurately control the output swing of the first stage. CLS-SA ensures robust control over this swing by supplying the SA current for the first stage from a pre-charged capacitor rather than directly from a normal supply. The problem of changes in offset voltage due to aging effects and temperature variations is also a less of an issue with CLS-SA because of the upsized transistors used for Big_SA and Small_SA for the given input swing.

References

N. Verma, A. Chandrakasan, A 256 kb 65 nm 8T sub threshold SRAM employing sense-amplifier redundancy. IEEE J. Solid-State Circuits **43**(1), 141–149 (2008)

M. Pelgrom et al., Matching properties of MOS transistors. IEEE J. Solid-State Circuits **24**(5), 1433–1439 (1989)

S. Cosemans, W. Dehaene, F. Catthoor, A 3.6pJ/Access 480 MHz, 128kbit on-chip SRAM with 850 MHz boost mode in 90 nm CMOS with tunable sense amplifiers. IEEE J. Solid State Circuits **44**(7), 2065–2077 (2009)

M. Bhargava, M. P. McCartney, A. Hoefler, K. Mai, Low-overhead, digital offset compensated, SRAM sense amplifiers, in *Proceedings of IEEE Custom Integrated Circuits Conference* (*CICC*), pp. 24-2-1-24-2-4, Sept 2009

M. Sinangil, N. Verma A. Chandrakasan, Reconfigurable 8T ultra- dynamic voltage scalable (U-DVS) SRAM in 65 nm CMOS. IEEE J. Solid-State Circuits **44**(11), 3163–3173 (2009)

S. Cosemans, Variability-aware design of low power SRAM memories, U.D.C 621.3.049.77, May 2009

V. Sharma et al., A 4.4pJ/Access 80 MHz, 2 K word X 64b memory with write masking feature and variability resilient multi-sized sense amplifier redundancy for W.S.Nodes, *in Proceedings of IEEE European Solid State Conference* (*ESSCIRC*), pp. 358–361 Sept 2010

A. Kawasumi et al., A low-supply-voltage-operation SRAM with HCI trimmed sense amplifiers. IEEE J. Solid State Circuits **45**(11), 2341–2347 (2010)

V. Sharma et al., 8T SRAM with mimicked negative bit-lines and charge limited sequential sense amplifier for wireless sensor nodes, *in Proceedings of IEEE European Solid State Circuits Conference* (*ESSCIRC*), pp. 531–534 Sept 2011

Chapter 7
Prototypes

7.1 Introduction

This chapter describes two prototypes of Static Random Access Memory (SRAM) macro. The test chips (IM_90 and IM_65) have been successfully developed, fabricated, and tested in order to validate the proposed low energy and variability resilient circuit techniques discussed in the previous chapters. First a design overview of IM_90 (first prototype) is provided followed by IM_65 (second prototype). This chapter concludes with the performance comparison of IM_90 and IM_65 with the current state-of-the-art for the wireless sensor node applications.

7.2 IM_90 (First Prototype 90 nm IP)

7.2.1 Target Application

The embedded memory design targets bio DSP chip, operating below 100 MHz of frequency range for wireless sensor nodes. Embedded memories consume a major proportion of the power budget of the low power sensor nodes applications (Nil et al. 2007); (Kwong et al. 2009). Software code optimization techniques (Verma and Marwedel 2007), tend to improve locality of data/instruction fetches. In memory hierarchy system, the largest memories have the least number of accesses per word, whereas the largest number of accesses per word is of L1 memory. Therefore, ultra low energy SRAMs for L1 data/instruction memory is a fundamental component of the wireless sensor node architecture, to meet the energy limitations of energy scavenging. Hence, energy efficient implementations for these small memories are a key requirement to enable further extensions of the capabilities of the energy scavenged wireless sensor nodes.

V. Sharma et al., *SRAM Design for Wireless Sensor Networks*,
Analog Circuits and Signal Processing, DOI: 10.1007/978-1-4614-4039-0_7,

Table 7.1 Memory dimensions and SRAM 6T cell details

Technology	90 nm LP, 3Vt CMOS
Word length	64 bits
Memory size	128 kbits, 2 K words
Cell type	SRAM 6T, HVT transistors
Cell size (logic DRC)	32.66F × 8.11F ($264.87F^2$)
Memory size	1.02 mm × 1.4 mm
SA calibration scheme	MS-SA-R with 2 options

7.2.2 Design Innovation Contributions

A low power design feature of IM_90 based on the divided word line decoder architecture with the low swing hierarchical bit-lines includes:

1. High Vt transistors based SRAM cells reduces the memory array leakage and enhances SRAM cell stability (See Figs. 2.2 and 2.3, in Chap. 2).
2. Innovative local assist circuitry (See Sect. 4.5.1, in Chap. 4) used during READ & WRITE operation lowers the energy consumption and also adds more variability resilience compared to the conventional local assist techniques.
3. The WRITE masking feature (See Sect. 5.2.5, in Chap. 5) further decreases the write energy by facilitating the partial WRITE operations.

The novel Multi-Sized SA redundancy (See Sect. 6.2.5, in Chap. 6) proposed for the global read sense amplifiers accommodates process variation and achieves an ultra low energy access for the given target yield.

7.2.3 Design Description

Table 7.1 shows the design details of IM_90. The designed 2K × 64 bits L1 SRAM (IM_90) consumes 4.4 pJ/access at 80 MHz and the leakage power is 0.6 uW at the retention mode. The power supply voltage (VDD) of the SRAM matrix is at 0.4 V. SRAM 6T cell design is based on the logic Design Rule Checks (DRC).

1. **Memory Floorplan**

The memory matrix consists of 512 cells by 256 cells. Figure 7.1 shows top level memory organization. The memory matrix is divided into 4 matrix columns. Each row of the memory matrix has its own global word line activation (GWL) signal. Each column has 64 word blocks and 64 pairs of vertical global bit-lines (VGBL). A common horizontal global bit-lines (HGBL) bus is shared by VGBL pairs of all the columns. The columns are enabled by the column select (CS) activation signals of a decoder. The CS activation signal also activates the interface MUX of VGBL pairs of the activated column with the HGBL bus. The word

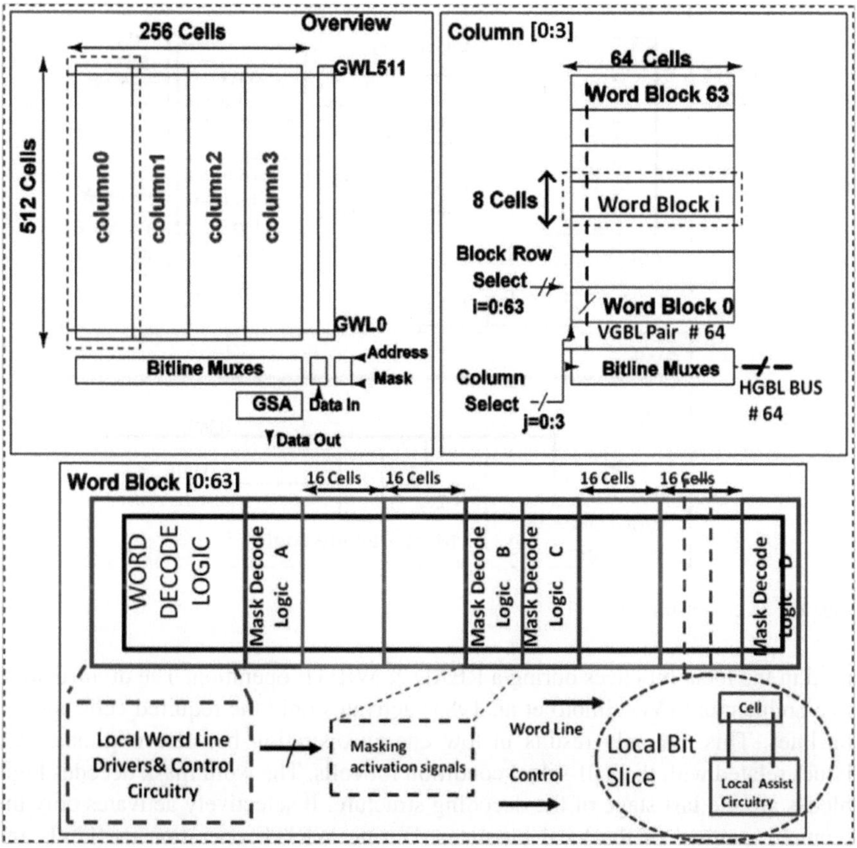

Fig.7.1 Memory organization

block consists of word decode logic, which performs the local decoding function and the mask decode logic blocks for decoding the masking information at the local level during the WRITE operation. Each mask decode logic block, generates controls for 16 local bit-slices.

2. **Decoder Structure**

The 11 address bits are decoded with 3 stage static AND–AND decoding logic. The decoding structure is shown in Fig. 7.2. The first stage of decoding logic generates 64 block row select (BRS) signals, 4 column select (CS) signals, and 8 within block row select (WBRS) signals. The 512 GWL activation signals are generated by the second stage of the decoding logic, which combines BRS and WBRS signals. The Word Block activation signal (Block_En) is generated from CS and the BRS signals. The Word Blocks are activated by Block_En. The word decode logic generates the local word line (LWL) by combining GWL and Block_En signals. It also generates activation signal for the local assist circuitry

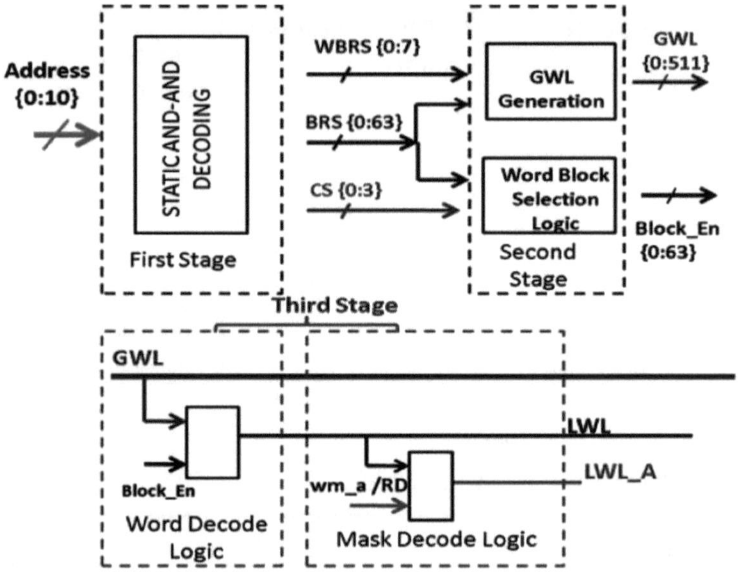

Fig. 7.2 Decoder structure

used in the local bit slices during a READ & WRITE operation. The divided word line architecture (Yoshimoto et al. 1983) activates only the required cells and the bit-lines. This not only results in low energy operation but also eliminates the issues related with the half-select condition for cells. The word mask decodes logic blocks are the last stage of the decoding structure. It selectively activates only the required quarters of the local bit-slices. During WRITE operation the LWL and other the control signals for the local assist circuitry are decoded to obtain a masked local word line and a masked activation signals.

3. **Hierarchical Bit-Lines and READ/WRITE Timing**

This design features hierarchical bit-lines with reduced voltage swing. This reduces the energy consumption associated with the charging and discharging of highly capacitive global bit-lines during the READ/WRITE operation. Figure 7.3 shows reduced swing bit-line hierarchy for energy efficient READ/WRITE operation along with timing waveforms at the global level. During a READ operation the HGBL pre-charge circuitry is activated which pre-charges the HGBL bus. Then the decoder activates the VGBL-HGBL interface MUX and disables the VGBL pre-charge circuitry. The word block decoder generates the required control signals for the local bit-slices. After the local block processing at the local bit-slice level, the READ information is transferred back onto the VGBL/HGBL bus. READ operation concludes with the disabling of the interface MUX and enabling the VGBL pre-charge circuitry. Similary during the WRITE operation the write drivers enable (WD_En) signal transfers the data input information onto the HGBL bus. Then the VGBL pre-charge circuitry of the selected matrix column is disabled

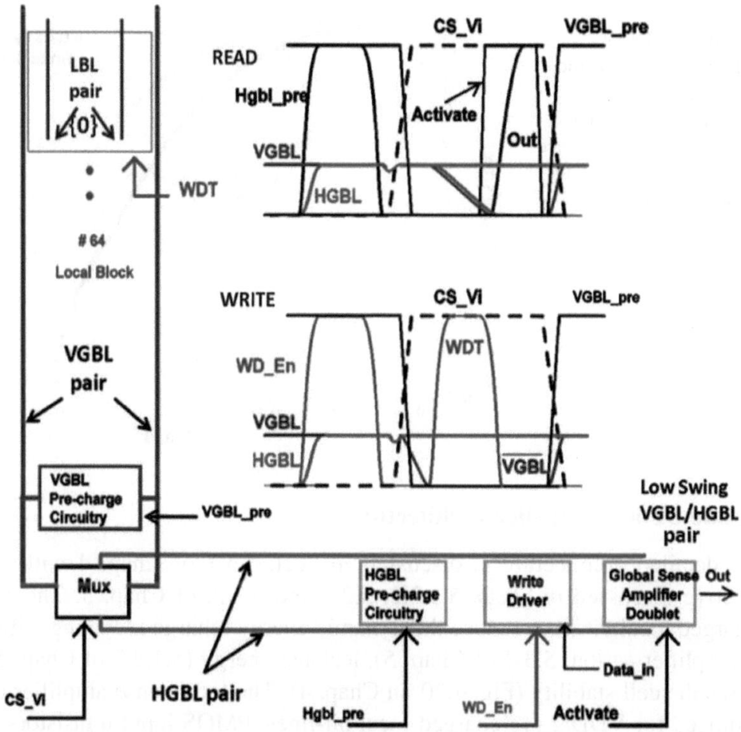

Fig. 7.3 READ/WRITE operation and bit-line hierarchy

and the low swing input data information is then made available to the VGBLs of the selected columns. This low swing data information is then transferred to the local bit-slices of the activated word block by the write data transfer (WDT) activation signal. Then the local block processing is done for the WRITE operation. The HGBL pre-charge circuitry and global sense amplifiers are not activated during WRITE operation. Similarly write drivers are not used during READ operation. Timing signals viz. HGBL pre-charge activation, WD_En and WDT are generated from the memory clock input and the write flag input signals.

4. **Enhanced cell stability**

The high Vt transistors based SRAM cell reduces leakage and read upset failures as discussed in (Figs. 2.2 and 2.3). The VDD/2 pre-charge value for the short local bit-lines further improves cell stability (Fig. 4.20). High Vt transistor based SRAM cells using dynamic read stability (short local bit-lines: 8 cells per local bit-line) coupled with reduced pre-charge voltage of VDD/2 for the local bit-lines ensure a very high cell stability for this design. Figure 7.4 shows impact of VDD/2 pre-charged local bit-lines for this prototype in improving read stability.

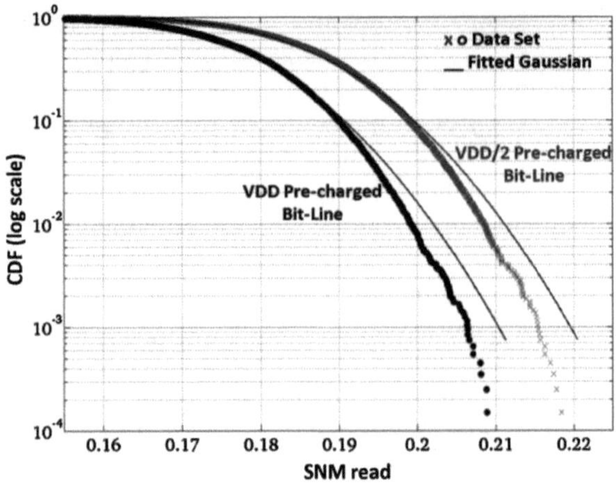

Fig. 7.4 Distribution of SNM read for VDD (conventional local bit-line architecture) versus VDD/2 (IM_90) pre-charge voltage for the local bit-lines obtained by performing 10 K Monte Carlo runs for VDD = 0.8 V

5. **Pseudo 8T local bit-slice architecture**

The detailed architecture is discussed in Sect. 4.5.1 of Chap. 4 with energy perspective discussed in Sects. 5.1.2.2, 5.2.5 and 5.3.2, of Chap. 5. The VDD/2 pre-charged local bit-lines reduce the dynamic energy (charge recycling with local sense amplifier action, 5.3.2 of Chap. 5), leakage energy (5.1.2.2 of Chap. 5) and increases the cell stability (Fig. 4.20, in Chap. 4). The local sense amplifier design is optimized for VDD/2 pre-charged local bit-lines. PMOS input transistors of the local sense amplifier are made stronger than the NMOS input transistors, Fig. 7.5. The delay contribution of the local assist circuitry (including SRAM cell and local sense amplifier) is only 20 % of the total SRAM macro access time. Therefore, the minor performance loss due to the VDD/2 pre-charged local bit-lines is not a major concern, especially considering the benefits achieved. And also the design targets wireless sensor nodes applications (Freq tens of MHz). Table 7.2, shows impact of VDD/2 pre-charged short local bit-lines.

6. **Low Swing Masking WRITE operation**

The detailed concept of low swing WRITE masking is discussed in Sect. 5.2.5, Chap. 5. The dynamic energy consumption of low swing WRITE operation is further reduced with the selective activation of circuits for the unmasked quarters of the word. The WRITE energy is reduced by 44 % when writing a ¾ masked word compared to the full word. The option of masking feature results in an overall decrease in energy consumption for the applications in which the number of partial writes **(Nw)** are relatively high compared to the number of reads **(Nr)**. For example, let us take an application in which the **(Nr/Nw)** is 1 **[Nw = 1/2Na, Nr = 1/2Na]**.

$\mathbf{E_{read,write}}$ Read, write energy consumption without masking.

$\mathbf{E_{mread,\ write}}$ Read, write energy consumption with masking.

$\mathbf{E_{mpwrite}}$ Write energy consumption for writing partial word for e.g. writing ¼ of word.

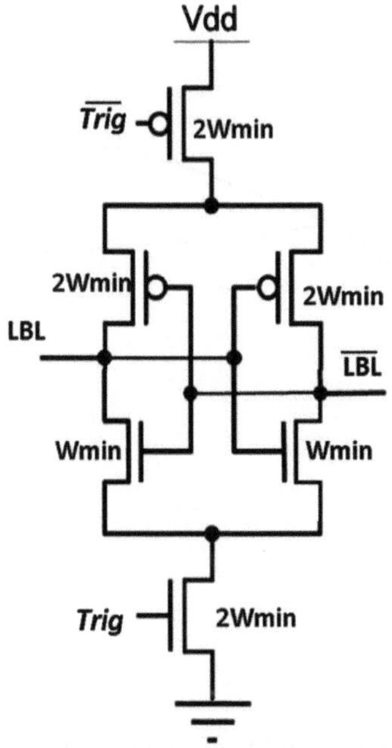

Local Bit-Line Pre-charge	σV*offset* LSA	μ delay* (ps) LSA
VDD	29.97mV	114.9
VDD/2	20.47mV	363.56

*delay: LSA triggered -90% signal rise time

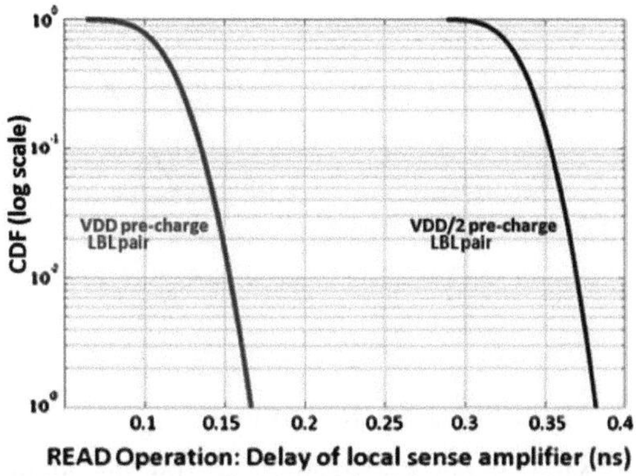

Fig. 7.5 Local sense amplifier sizing and the impact of local bit-line pre-charge voltage. Delay of local sense amplifier is based on 1 K Monte Carlo runs

Table 7.2 Impact of VDD/2 pre-charge value for the short local bit-lines

Active energy	Reduces
Leakage energy	Reduces
Cell stability (SNM read)	Improves
LSA offset	Reduces*
Delay of local assist circuit	Degrades**

* local sense amplifier design optimized for VDD/2
** total impact of this delay on overall performance is minimal

Na Total number of memory accesses {value taken 30}.
$\mathbf{E_{tm}}$ (with masking) Nr × Emread + Nw × Emwrite.
$\mathbf{E_{tnm}}$ (without masking) Nr × Eread + Nw × Ewrite.

The additional decode circuitry required for the selective activation of the assist circuitry for the low energy implementation of the masking feature increases the read energy consumption by 16 %. In applications, where the partial write accesses are in good proportion to the read accesses it reduces the overall access energy consumption. Figure 7.6 shows the measured total energy consumption for 30 memory accesses for different proportions of READ/WRITE accesses.

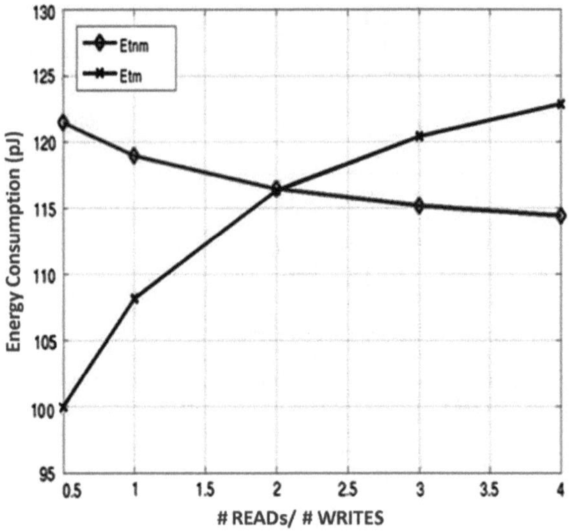

Fig. 7.6 Total energy consumption based on the measurement results of test chip $E_{tnm} = 118.94$ pJ and $E_{tm} = 108.15$ pJ for Na = 30 & # READs (Nr)/# WRITES (Nw) = 1

7. **Multi-Sized SA Redundancy**

Multi-Sized SA Redundancy for the global sense amplifiers enables energy efficient low swing sensing operation. The concept and the design implementation details are discussed in Sect. 6.2.5 of Chap. 6.

7.2.4 Measurement Results

The proposed memory is fabricated in a 90 nm LP CMOS process. Figure 7.7 shows die photograph of proposed memory. The prototype consists of an SRAM macro and test circuitry. The test circuitry consists of an input shift register, an output shift register, and delay measurement circuitry. The write data, write flag information, masking bits, and address bits are shifted serially into an input shift register. The output of SRAM macro is loaded into the output shift register and then shifted serially out. The delay measurement circuitry consists of latch lines to monitor internal signals of SRAM macro. Both the SA output and the memory activation signal are monitored with such latch line. The memory access time is the difference between the memory activation time and the time at which the SA output changes. For the energy consumption a random data pattern is taken. The charge calculation per access is done by integrating the current derived from the VDD supply line for the access time duration.

Figure 7.8 shows average measurement results for the tested sample chips at 25 °C and at VDD = 0.8 V. The memory operates at 80 MHz consuming 4.42 pJ/access for READ operation, 5.02 pJ/access for WRITE operation, and 2.79 pJ/access for ¾ masked WRITE operations. The active leakage power is 5.26 uW and

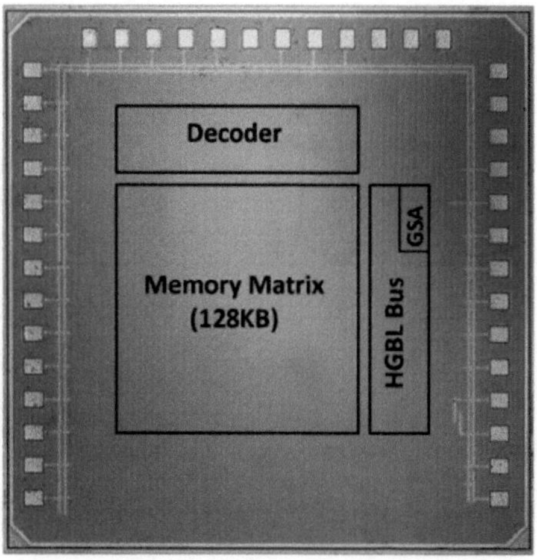

Fig. 7.7 Die photograph

Maximum Frequency [MHz]	80 MHz
Active Energy per Access for READ with all 64 2sigma SA	4.39 pJ
Active Energy per Access for READ with all 64 6sigma SA	5.11 pJ
Active Energy per Access for READ #61 2σ SA & #3 6σ SA	4.42 pJ
Active Energy per Access for WRITE (no masking)	5.02 pJ
Active Energy per Access for WRITE (3/4 masking)	2.79 pJ
Active Leakage Power	5.26 uW
Static Leakage (retention at 400mV) Power	0.6 uW

Fig. 7.8 Measurement results at 25 °C and VDD = 0.8 V

the static leakage power at retention mode is 0.6 uW when the data is held at 0.4 V. Figure 7.9 compares the energy breakdown of READ operation of SRAM macro with 6σ SA and with MS-SA-R, N = 2 {#61 2σ SA and #3 6σ SA}.

7.3 IM_65 (Second Prototype 65 nm LP)

7.3.1 Target Application

The classic design paradigms in order to meet the challenges posed by the technology scaling rely on the upsizing and on the extra design margins. This over sizing and too much insertion of the design margins result in an excessive degradation of the energy consumption and performance. Therefore, the circuit

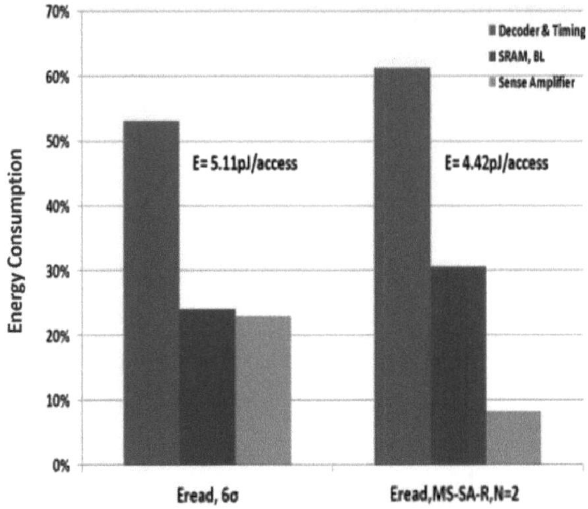

Fig. 7.9 Energy breakdown of READ operation VDD = 0.8 V and 25 °C with 6σ SA (traditional, # of SA = 64) versus MS-SA-R N = 2 {# of 2σ SA = 61 & # of 6σ = 3}

design techniques which improve the operating margins of SRAM without increasing the energy consumption are required. The second prototype of SRAM macro (IM_65) is also designed for the wireless sensor node applications (<100 MHz). The difference between IM_65 and IM_90 is that IM_65 also addresses the issue of increasing the operating margins of SRAM for the advanced technology nodes. The key differences between IM_65 & IM_90 are as follows:

1) The SRAM core of IM_65 is based on 8T SRAM cell structure compared to 6T SRAM cell used for IM_90.
2) The sequential voltage optimization technique (MNBL) proposed for the WRITE mechanism in IM_65 is far more superior in performance and solves the issues related with the conventional voltage optimization technique like VDD lowering, used in IM_90.
3) The sense amplifier used in IM_65 does not rely on calibration in order to solve the energy-offset tradeoff issue compared to the calibration based solution (Multi-sized SA redundancy) used in IM_90.

7.3.2 Design Innovation Contributions

The design innovations in IM_65 prototype are:

1) Reduced swing dual Vt 8T SRAM cell (Sect. 4.c, of Chap. 2 Sect. 5.1.2.4 of Chap. 5) reduces the static leakage and aids in improving the variability resilience.
2) A proposed write method, Mimicked Negative Bit-line technique (Sect. 3.5.3 of Chap. 3), improves the write margin and resolves the issues related with the existing negative bit-line technique.

Table 7.3 Memory details of IM_65

Technology	65 nm LP, 3Vt CMOS
Memory size	64 kbits, 1 K words
Cell type	Dual Vt 8T SRAM (HVT for 6T part + SVT for the 2T read buffer)
Cell size	1.6 um^2 (logic DRC)
Decoder architecture	Static 3-stage, fully sub-divided word line
Sense amplifier	Charge limited sequential sense amplifier (CLS-SA)
Bit-line structure	Hierarchical static low swing write bit-lines and low swing read bit-lines

3) Low Swing Static WRITE mechanism (Sect. 5.2.6 of Chap. 5) achieves an ultra low energy WRITE operation.
4) A novel calibration free Charge Limited Sequential sense amplifier (Sect. 6.3 of Chap. 6) enables low swing sensing and improves the energy-offset tradeoff. It also avoids the additional increase in the memory test costs and test time, associated with the calibration based techniques.

7.3.3 Design Description

Table 7.3, shows design details of IM_65. A 64 kbit embedded SRAM macro in 65 nm LP CMOS achieves an energy consumption of 2.65 pJ/access at 90 MHz. The design innovations improve the variability resilience and achieve low energy consumption. Reduced swing dual Vt 8T SRAM cell mitigates leakage by 7x. Write-ability is improved by the Mimicked Negative Bit-line technique, which reduces write failures by 10^3x. The energy consumption is further reduced by using a novel Charge Limited Sequential sense amplifier, which achieves a σVoffset of 14 mV with energy consumption of only 11 fJ/decision, without requiring post silicon tuning.

1. **Memory Floorplan**

The memory matrix consisting of 256 × 256 cells is organized into 4 columns (Fig. 7.10). The matrix column is further divided into 16 local blocks. The local block consists of a local decode logic and 64 local bit-slices. A local bit-slice consists of 16 8T SRAM cells and a write receiver. This design uses hierarchical divided write bit-lines. However, there is no hierarchy for the read bit-lines. The write ports of the 16 8T SRAM cells on the local write bit-lines shares a local write receiver. The low swing signal from the global write bit-lines is amplified to the full swing on the local write bit-lines. The design also uses extended bit-lines and enables a set of global sense amplifiers to be sufficient for the entire memory matrix.

The 10 address bits are decoded with the static AND–AND 3 stage decoding logic. The first stage of decoder converts the address bits into 16 block select signals (activates a row containing local blocks), 4 column selects (indicates column activation), and 16 within block row selects (indicates a word within the

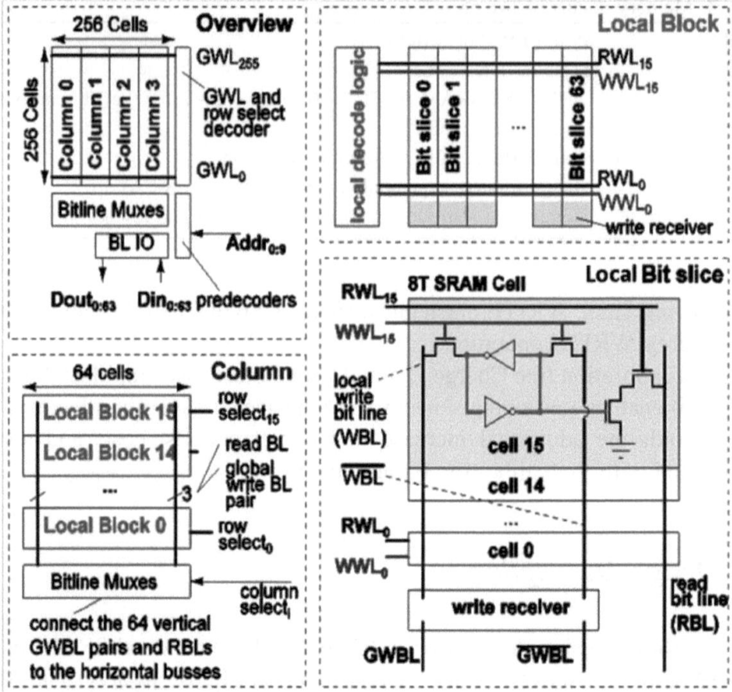

Fig. 7.10 Memory organization

local block). There are two groups for the second stage decoding. One group decodes the block row select signals with the column select signals and activates the local block. The second group decodes 16 block row select signals and 16 within block row select signals into 256 global word lines. The last stage of decoding generates within local block control signals from the (GWLs) and the local block signals.

The GWLs are combined with the column selects to activate the fully subdivided local read and write word line. The divided word line architecture for READ and WRITE operation activates only the required cells and the bit-lines. This results in energy savings and avoids the half select condition.

2. **Reduced Swing Dual Vt 8T SRAM cell**

Reduced swing Dual Vt 8T SRAM cell reduces leakage by 7 × and increases variability resilience as discussed in Sect. 4.c, Sect. 5.1.2.4 of Chap. 5.

3. **Mimicked Negative Bit-line Technique**

Overcomes the limitation associated with the conventional negative bit-line technique. The concept and details are provided in Sect. 3.5.3, Chap. 3.

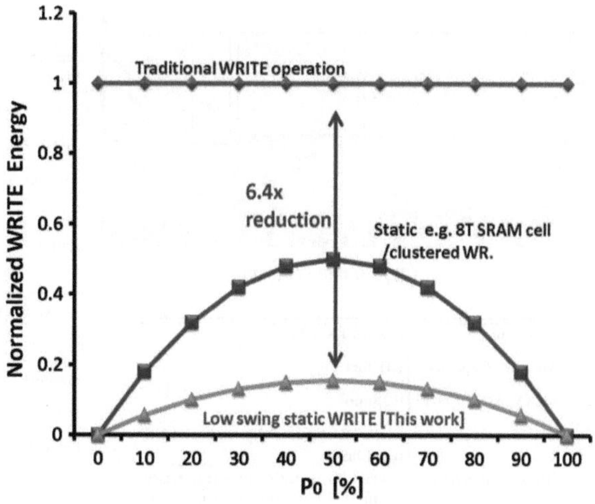

Fig. 7.11 Energy consumption comparison of low swing static WRITE bit-lines with the traditional WRITE operation (pre-charged complementary bit-lines) and static complementary bit-lines

4. **Low Swing Static WRITE Operation**

In traditional SRAM design, the bit-lines are shared for the READ and WRITE operations The READ and WRITE operations occur in an interleaved fashion. The energy is consumed during each cycle, independent of the data pattern because the bit-lines must be pre-charged for the READ operations. The separate read and write bit-lines with write bit-lines does not require pre-charging (static write signals) helps in reducing WRITE energy consumption (Static).

Static WRITE operation as discussed in Sect. 5.2.6 of Chap. 5coupled with low swing bit-lines results in an extremely low energy WRITE operation (Fig. 7.11).

5. **Charge Limited Sequential Sense Amplifier (CLS-SA)**

Charge limited sequential sense amplifier enables low swing sensing and improves the energy-offset tradeoff. It is a calibration free solution for energy-offset tradeoff with the result the additional increase in the memory test cost and test time as associated with the calibration based solution is avoided. The concept and design details of CLS-SA are discussed in Sect. 6.3 of Chap. 6.

7.3.4 Measurement Results

The prototype consists of an SRAM macro and test circuitry, is fabricated in a 65 nm LP CMOS process (Fig. 7.12). The on-chip test circuitry consists of 2-entry input shift registers, 2-entry output shift registers, and a delay measurement circuit. The on-chip test circuitry allows performing 2 random accesses at speed, either once or in a repeated loop. This allows testing the relevant data and access patterns. The delay measurement circuitry comprises of latch lines and is used to monitor internal signals of SRAM macro. The memory operates at 90 MHz

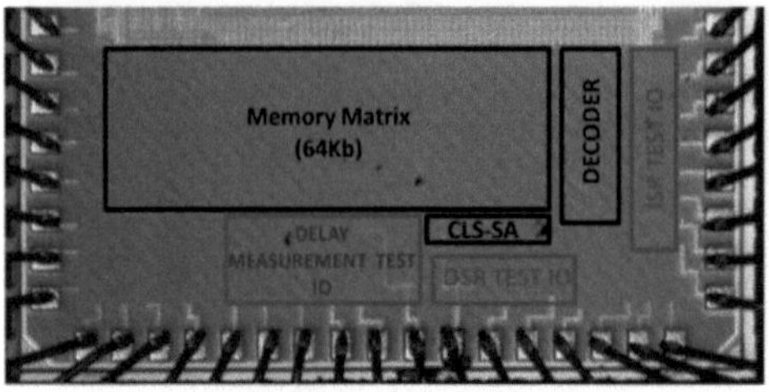

(a) Chip photograph.

Technology	65nm LP CMOS
Memory Capacity	64Kbits
SRAM Macro Size	$0.38mm^2$
Decoder Structure	Static 3-stage, fully sub-divided word lines
Bit-line Structure	Hierarchical static low swing write bit-lines & low swing read bit-lines
Cell Type	Read decoupled 8T (dual Vt, HVT for 6T part & SVT for read buffer Tr)

(b) Table I. Test chip details.

Frequency	90MHz
Active Energy READ all "1s"	2.83pJ/access
Active Energy READ all "0s"	2.68pJ/access
Average READ Energy "1s/0s"	2.76pJ/access
Active Energy WRITE all "1s"	2.5pJ/access
Active Energy WRITE all "0s"	2.51pJ/access
Standby Power Consumption Worst case (all "1s" stored)	0.084uW
Performance of 1 CLS-SA	11.36fJ/decision σVoffset = 14.3mV leakage= 134nW

(c) Table II. Measurement Results at 20°C & VDD= 0.8V.

Fig. 7.12 Die photograph measurement results

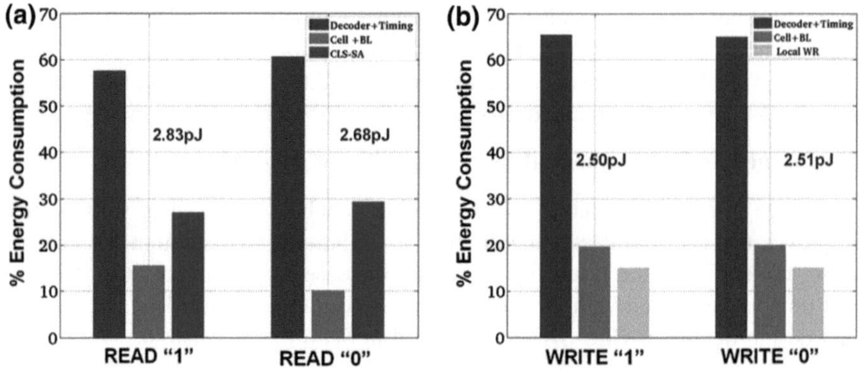

Fig. 7.13 Energy breakdown of READ and WRITE operation at VDD = 0.8 V and 20 °C

(Fig. 7.12c) and consumes 2.76 pJ/access for reading word with 50 % "1 s" and 50 % "0 s". The write energy consumption for words with all "1 s" or all "0 s" is 2.5 pJ/2.51 pJ per access at VDD = 0.8 V at 20 °C. Figure 7.13 shows the energy breakdown of SRAM macro for READ and WRITE operation.

The fabricated chips are fully functional (READ/WRITE) for the temperature range for −20 °C – 70 °C at VDD = 0.8 V. For the retention test data is written to all the cells of the memory matrix using a known good supply voltage for the cell. Then the cell supply voltage is temporarily reduced to some trail value. Finally, the cell content is read out using a known good supply voltage for the cell. This is repeated by decreasing the value of trial voltage. Figure 7.14 shows the measured

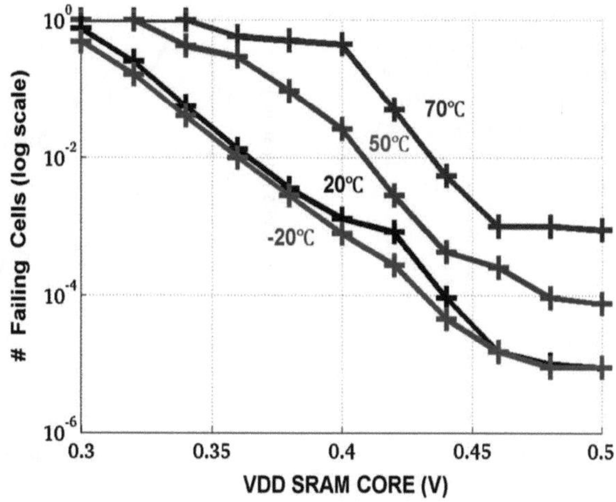

Fig. 7.14 Measured fraction of cells (# failing bit cells/# total bit cells in prototype) that fail to hold data at reduced VDD for different temperatures

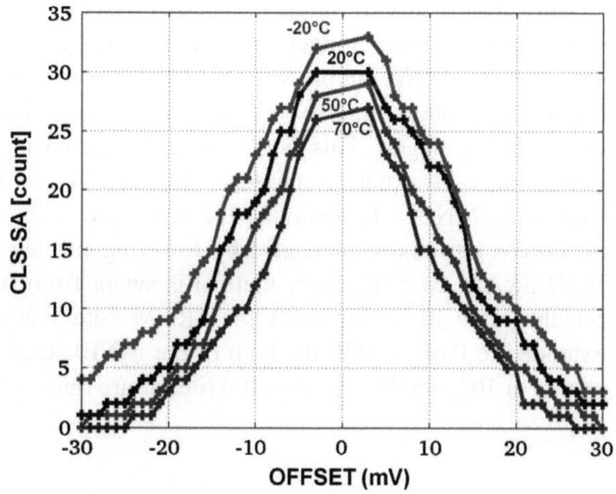

Fig. 7.15 Offset measurement of charge limited sequential-SA for different temperatures

cumulative distribution of cell retention margins for all 64 K cells on a die. There are no retention failures for the VDD SRAM core at 0.47 V at 20 °C (room temperature). The worst case (all "1 s" stored) static leakage power consumption is 0.084 uW at 20°C and 0.47 V, 0.066 uW at -20 °C and 0.46 V and 0.608 uW at 70 °C and 0.56 V. For 70 °C, the retention voltage has to be increased to 0.56 V.

The CLS-SA consumes 11.36 fJ/decision for a σV_{offset} of 14.3 mV (Fig. 7.12c). The CLS-SA improves the tradeoff between energy and SA offset, without the overhead of post-silicon tuning. The standard deviation of the measured offset voltage σV_{offset} of the 64 CLS-SAs on a single die is 14.3 mV. Figure 7.15 shows offset measurements for different temperatures. The output swing of Big_SA available to the Small_SA increases at higher temperatures because of the lower Vt of Mtop and the cross-coupled PMOS transistors. The count of CLS-SA failures increases at lower temperature.

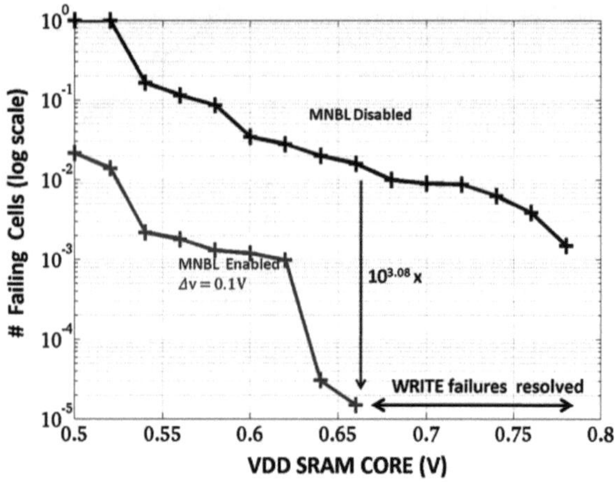

Fig. 7.16 Measured fraction of cells that cannot be written correctly at −20 °C for reduced VDD. MNBL technique ($\Delta v = 0.1$ V) results in $10^3 \times$ reduction in the cell failure rate

The write ability of the SRAM cell is measured as the lowest supply voltage at which the cell can no longer be written correctly. A zero is written to all the cells. The memory matrix supply voltage is lowered to a trial voltage and a write access is performed attempting to write one to the cells. Then the cell state is examined using safe voltages. This is repeated for increasing the value of trial voltage. Furthermore, in order to measure the effectiveness of the proposed write assist technique (MNBL) in resolving the cell write failures. The number of write failures were artificially increased by lowering the temperature (−20 °C) and then VDD SRAM core (memory matrix) is swept from 0.8 to 0.5 V. The write functionality at −20 °C for the VDD SRAM core voltage of 0.78 V can be further extended to 0.66 V with the help of the MNBL technique. The MNBL technique results in $10^3 \times$ reduction in cell write failure rates for VDD = 0.66 V (Fig. 7.16).

7.4 Comparison with the State-of-the-Art

Table 7.5 compares the performance of the proposed test chips (IM_90 Sharma et al. (2010) and IM_65 Sharma et al. (2011)) with other state-of-the-art SRAM macros targeting sensor node applications. To ease the comparison, a figure of merit (FOM) was added to the table: the energy divided by the number of bits per word. Although this FOM favors the SRAM macros with longer word length, the energy per access per bit of this SRAM macro is 3.4–21 times lower than that of the state-of-the art memories (Fig. 7.17).

The area of the SRAM macro is rather larger, primarily due to the use of local assist circuitry, 8T cells, and due to the unavailability of litho optimized cells for academic purposes. The proposed circuit design techniques target low energy consumption. The tradeoff of integrated circuit (SRAM macro) area for achieving lower energy consumption is a viable option for reducing the overall form factor of

Table 7.4 Comparison with other state-of-the-art designs

Design	Technology node	Operating VDD	Capacity	Word length (bits)	Freq MHz	Capacity/Area[1] Mb/mm2	Energy
Takeda et al. (2011)	40 nm	(1.0–1.2) V	2 Mb	128	250	1.834	110pJ@1.1 V
Sinangil et al. (2011)	28 nm	(0.6–1.0) V	128 kbit	256	(20–400)	1.97	140pJ@0.6 V
Yoshimoto et al. (2011)	40 nm	(0.5–0.8) V	512 kbit	.....	(0.625–22)	0.431	8.8pJ@0.5 V
Sharma et al. (2011)	65 nm	0.8 V	64 kbit	128	90	0.168	2.65pJ
Sharma et al. (2010)	90 nm	0.8 V	128 kbit	64	80	0.09	4.42pJ
Kushida et al. (2009)	65 nm	0.9 V	256 kbit	256	127	0.8	36pJ
Sinangil et al. (2009)	65 nm	(0.4–1.2) V	64 kbit	128	(0.02–200)	0.046	23pJ@0.8 V
Kwong et al. (2009)	65 nm	(0.3–0.6) V	128 kbit	64	(0.0087–1)	0.094	27.2pJ@0.5 V
Verma et al. (2008)	65 nm	(0.35–0.5)V	256 kbit	128	(0.03-1)	0.12	30pJ@0.5 V
Takeda et al. (2006)	90 nm	(0.5-1.0) V	64 kbit	16	(50–833)	0.23	3.2pJ@0.5 V

[1] Area converted from other technology nodes to 65 nm except for designs in 90 nm

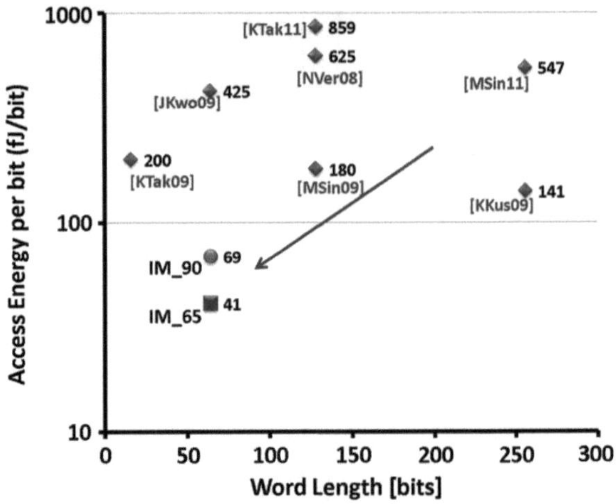

Fig. 7.17 Comparison of access energy per bit with the state-of-the-art memories

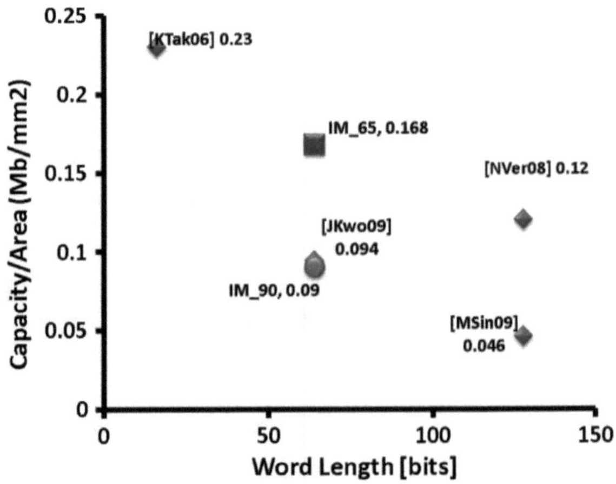

Fig. 7.18 Comparison of capacity/area with the state-of-the-art academic memories (litho optimized designs not considered for fair comparison)

the WSN nodes. The increased energy efficiency of the integrated circuits (SRAM macro) would require smaller sized batteries. But the area overhead of the proposed designs is comparable to the state-of-the-art academic designs not relying on litho optimized SRAM cells (Fig. 7.18).

The SRAM L1 memory fabricated in 90 nm LP (IM_90) discussed in Sect. 7.2 features ultra low power variability resilient circuit techniques. The local assist circuitry includes a local sense amplifier on the short local bit-lines and a gated read buffer. The local sense amplifier reduces the impact of the cell read current on access speed, which allows minimum sized high Vt cell transistors, reducing leakage. It also enables charge re-cycling with VDD/2 pre-charged short local bit-lines. The use of gated read buffer enables pseudo 8T SRAM cell type READ operation with 6T SRAM cell and also eliminates the bit-line leakage under idle conditions. The sense amplifier used in local bit-line architecture also serves as a write receiver during WRITE operation, saving area and reducing leakage. Multi-

Sized SA redundancy (MS-SA-R) reduces the energy consumption of the global read sense amplifiers as compared to the traditional calibration based schemes. Measurement results show that 128 kbit 6T SRAM 90 nm LP CMOS consumes 4.4 pJ/access when operating at 80 MHz and 0.8 V.

The 64 kb SRAM macro (IM_65) discussed in Sect. 7.3 features a RSDVt 8T SRAM cell based core which enhances Iread, SNMread and reduces leakage. The reduced swing read bit-lines with dual Vt transistors for the SRAM cell achieve 7x (2 × due to dual Vt and 3.5x due to reduced swing bit-lines) reduction in leakage current for the worst case stored data pattern. Hierarchical low swing static write bit-lines results in 6.4x reduction in the dynamic energy consumption. The proposed Mimicked negative bit-line technique result in 10^3 x reduction in the write failures and also avoids the potential risk of latch up with the existing negative bit-line technique. The novel sense amplifier circuit offers a SoC friendly solution. There is no requirement for post-silicon tuning with CLS-SA, as required by the calibration based SA design techniques. It solves the energy-offset tradeoff issue and enables the low swing read bit-line sensing. The dynamic energy consumption is reduced by 2.5x compared to the traditional low swing sense amplifier. The proposed SRAM macro test chip for wireless sensor nodes in 65 nm LP process achieves very low energy consumption. The average energy consumption (read/write) of the test chip operating at 90 MHz is 2.65 pJ/access at VDD = 0.8 V.

References

K. Kushida et al., A 0.7V single-supply SRAM with 0.495 um^2 cell in 65nm technology utilizing self-write-back sense amplifier and cascaded bit line scheme. *IEEE J. Solid-State Circuits* **44**(4), 1192–1198 (2009)

J. Kwong et al., A 65 nm sub-vt microcontroller with integrated SRAM and switched capacitor DC–DC Converter. *IEEE J Solid-State Circuits* **44**(1) 115–126 (2009)

M.D. Nil et al., Ultra low power ASIP design for wireless sensor nodes. *IEEE Conference on Electronics, Circuits and Systems (ICECS)*, pp. 1352–1355 (2007)

V. Sharma et al., A 4.4pJ/Access 80 MHz, 2 K word X 64b memory with write masking feature and variability resilient multi-sized sense amplifier redundancy for W.S.Nodes, in *Proceedings of IEEE European Solid State Conference (ESSCIRC)*, pp. 358–361 (2010)

V. Sharma et al., 8T SRAM with mimicked negative bit-lines and charge limited sequential sense amplifier for wireless sensor nodes. in *Proceedings of IEEE European Solid State Circuits Conference (ESSCIRC)*, pp. 531–534, Sept 2011

M.E. Sinangil, N. Verma, A.P. Chandrakasan, A reconfigurable 8T Ultra-Dynamic Voltage Scalable (U-DVS) SRAM in 65nm CMOS. *IEEE J Solid-State Circuits* **44**(11), 3163–3173 (2009)

M. Sinangil, H. Mair, A. Chandrakasan, A 28 nm high-density 6T SRAM wth optimized peripheral-assst circuits for operation down to 0.6V. *IEEE International Solid-State Circuits Conference (ISSCC)*, pp. 260–261, Feb 2011

K. Takeda et al., A read-static-noise-margin-free SRAM cell for low-VDD and high-speed applications. *IEEE J Solid-State Circuits* **41**(1), 113–121 (2006)

K. Takeda et al., Multi-step word-line control technology in hierarchical cell architecture for scaled-down high-density SRAMs. *IEEE J Solid-State Circuits* **46**(4), 806–814, (2011)

M. Verma, P. Marwedel, Advance memory optimization techniques for low-power embedded processors. ISBN 978-1-4020-5896-7, (Springer, Netherlands, 2007)

N. Verma, A.P. Chandrakasan, A 256 kb 65 nm 8T subthreshold SRAM employing sense-amplifier redundancy. *IEEE J. Solid-State Circuits* 141–149 (2008)

M. Yoshimoto et al., A 64 Kb full CMOS RAM with divided word line structure. *Proceedings of IEEE International Solid State Circuits Conference (ISSCC)* pp. 58–59 (1983)

S. Yoshimoto et al. A 40-nm 0.5V 20.1 uW/MHz 8T SRAM with low-energy disturb mitigation scheme. 2011 Symposium on VLSI Circuits, pp. 72–73, June 2011

Chapter 8
Conclusions

8.1 Synopsys of Contribution

The contributions for the design of low energy variability resilient Static Random Access Memory (SRAM) design are depicted in the flow chart below Fig. 8.1

SRAM Energy Reduction Contributions:

1) **Pseudo 8T architecture:** The local assist circuitry includes a local sense amplifier on the short local bit-lines and a gated read buffer. The local sense amplifier reduces the impact of the cell read current on access speed, which allows minimum sized high threshold voltage (Vt) cell transistors, reducing leakage. It also enables charge re-cycling with VDD/2 pre-charged short local bit-lines. The High Vt transistor based SRAM array with gated read buffer reduces leakage by 2.8x in 40 nm LP technology node. The use of gated read buffer enables pseudo 8T SRAM cell type READ operation with 6T SRAM cell and also eliminates the bit-line leakage under idle conditions.
2) **HBS bit-lines:** Hierarchical buffered segmented bit-line provides an energy efficient and high performance interface with very high cell stability without resorting to an energy expensive WRITE after READ mechanism. It solves the issues associated with SRAM design in the advance sub-nanometric technologies viz. access time degradation and increased power consumption. The VDD/2 pre-charged bit-lines, possible because of the segmentation done by the segment transistors. This reduces the power consumption. The access speed is increased by the use of segment transistors, driving the pre-charged read buffers. The power consumption reduction is 22 % less and the access speed is (1.3–1.4)x better compared to the state-of-the-art high speed SRAM designs [LChang08 and SIsh08].
3) **Low Swing WRITE with WRITE masking:** The bit-line voltage scaling helps in reducing the energy consumption and WRITE masking further aids in reducing the energy consumption. The WRITE masking acts as an energy

V. Sharma et al., *SRAM Design for Wireless Sensor Networks*, Analog Circuits and Signal Processing, DOI: 10.1007/978-1-4614-4039-0_8,

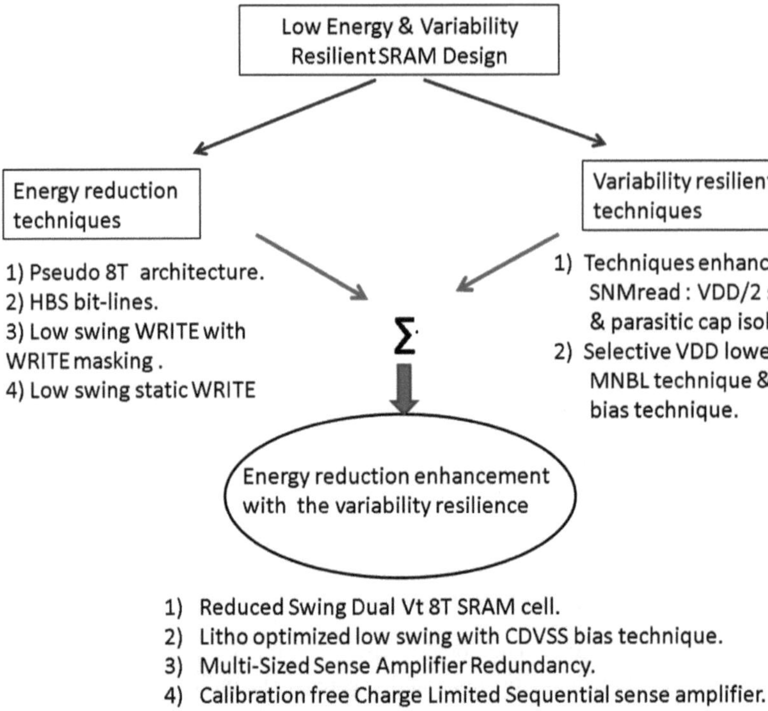

Fig. 8.1 Synopsys of Contributions

control knob feature by selectively masking the WRITE operation for certain bits of the data word length. The data correlation is exploited and the write operation for the certain set of bits of the data word is prevented. For every masked quarter the energy reduction gain can be between 25 % and 75 % compared to the low swing bit-lines based techniques.

4) **Low swing static WRITE:** The conventional bit-lines require pre-charging for the READ operation during each cycle, independent of the data pattern. This pre-charging results in huge energy consumption which can be avoided by utilizing static bit-lines (separated WRITE bit-lines from the READ bit-lines). The static write bit-lines with low swing signals for highly capacitive global bit-lines and signal amplification on the static local write bit-lines achieve an ultra low energy WRITE operation. The energy reduction gains with the static low swing write bit-lines is approximately 6.4x compared to the traditional WRITE operation.

SRAM Variability Resilience Contributions:

5) **VDD/2 pre–charged short local bit-lines:** The reduced local bit-line capacitance with only 8 SRAM cells result in enhanced improvement in the dynamic Static Noise Margin (SNM) read. The HVT based SRAM cells also offer higher

cell stability (read SNM). The dynamic SNM is further improved by reducing the noise source (bit-line charge) by decreasing the local bit-line pre-charge voltage (VDD/2). The VDD/2 pre-charged short local bit-lines result in 13 % and 20 % improvement in the read SNM for 65 nm and 40 nm technology node.

6) **VDD/2 pre-charged short local bit-lines and segment buffers:** VDD/2 pre-charged short bit-lines and parasitic bit-line isolation with the segment buffers enhance the SNM read. The segment transistors inserted in the local bit-line architecture isolate the parasitic capacitance of the local assist circuitry. The segment transistors are enabled only after the local bit-line has been discharged to a predefined limit and the word line signal has been disabled. The reduced pre-charged local bit-lines (VDD/2) further reduces the magnitude of the noise source thereby resulting in an enhanced SNM read. The dynamic SNM is improved by approximately 23 % compared to the VDD pre-charged local bit-lines with no parasitic capacitance isolation.
7) **MNBL technique:** It provides an alternative mechanism to the conventional WRITE assist methods with reduced constraints. There is no degradation of read SNM of unselected cells because of the voltage optimization which strengthens the PMOS devices. There is no data retention issue for the unselected cells on the activated column as the voltage difference between the VDD and VSS remains the same. The performance of the MNBL technique is comparable to the conventional negative bit-line technique. Also the potential risk of forward biasing the PN junctions as present with the conventional negative bit-line method is avoided. The MNBL technique results in 10^3x reduction in the SRAM cell write failures at VDD = 0.66 V, −20 °C, and 10^3x reduction at VDD = 0.5 V in 65 nm LP technology at room temperature.
8) **CDVSS bias technique: Differential VSS biasing is applied to reduce the mismatch offset of the non strobed local write receiver.** The differential VSS bias applied on the ground rails of local write receiver bipartite into two write assist techniques viz. the selective VSS raising and the negative bit-line mechanism for the accessed SRAM cell. The positive VSS bias applied weakens the pull up PMOS transistor of "H" side of the SRAM cell thereby improving write-ability of the accessed SRAM cell. The negative bias applied on the complement VSS has two advantages. First it makes the rise time faster during the WRITE operation thereby improving the write access time. Second it pulls the bit-line below 0V and generates the negative bit-line for the accessed SRAM cell without any extra added cost. The selective VSS raising and negative bit-line mechanism increases the SRAM cell write-ability. The probability of write failure for the worst corner is reduced by the factor of 10^3x at the scaled VDD levels (VDD = 0.55 V).

Energy reduction enhancement with variability resilience:

9) **RSDVt 8T SRAM cell:** The RSDVt 8T SRAM cell solves the leakage issue associated with conventional read decoupled 8T cell. The reduced swing read bit-lines with dual Vt transistors for the SRAM cell achieve 7x (2x due to dual Vt and 3.5x due to reduced swing bit-lines) reduction in leakage current for the

worst case stored data pattern (Q = 'H'). RSDVt 8T SRAM results in 26x reduction in the energy consumption compared to the conventional 6T SRAM cell in a 40 nm LP technology node.

10) **Litho Optimized local architecture with CDVSS Scheme:** Litho optimized local architecture reduces the transistor count and timing complexity associated with the conventional local assist circuitry. Reduced timing complexity, transistor count, and low VDD operation possible because of the applied CDVSS scheme reduces the energy consumption compared to the conventional hierarchical divided bit-line architectures. The area overhead of this solution is only 9 % compared to 38 % with the existing solutions. The reduced area overhead also aids in reducing the energy consumption. The high level of physical regularity in the layout of the local assist circuitry permits litho optimization thereby eliminating the memory matrix subarray design complexity associated with the conventional local assist circuitry. Thus the proposed circuit techniques promise the best area-energy-performance optimization compared to the existing solutions.
11) **Multi-Sized (MS)-Sense Amplifier (SA)-Redundancy for sense amplifiers:** The innovative Multi-Sized SA redundancy (MS-SA-R) calibration technique for the read sense amplifiers of the SRAM adds to the variability resilience and yields maximum energy reduction compared with existing calibration techniques. Compared to a traditional SA without calibration designed for the same differential input signal and the same yield, 2-fold MS-SA-R reduces the SA energy with a factor of 7, which is significantly better than the factor 2.2 of 2-fold SA-R and than the factor 4 of 2-fold SA tuning.
12) **Calibration free Charge Limited Sequential Sense Amplifier:** The novel Charge Limited Sequential (CLS) sense amplifier circuit offers a SoC friendly solution. There is no requirement for post-silicon tuning with CLS-SA, as required by the calibration based SA design techniques. It solves the energy-offset trade off issue and enables the low swing read bit-line sensing. The dynamic energy consumption is 26 % further less compared to the MS-SA-R.

8.2 Technology Scaling Perspective

This work describes several techniques which were used to realize a low standby power, low active energy memory that operates in the range of tens of MHz in 90 nm and 65 nm LP technology. This section discusses the technology scaling perspective of the proposed circuit design techniques.

8T cell with high-Vt 6T core: The 8T cell is a logical choice for designs in advanced technologies as it avoids read disturbs and allows optimizing the 6T core for write-ability. As the 6T core has no impact on memory speed, it can be implemented with slow, low leakage transistors, significantly reducing the standby power consumption.

Low-Vt read buffer with low-swing read: The read buffer current has a large impact on the memory speed, the use of fast, low-Vt transistors is recommended.

This not only improves the nominal read current, but also the variations on the read current thanks to the increased gate-source voltage overdrive. This improvement is most welcome in scaled designs with lower VDD and higher transistor variations. Additionally, the low pre-charge voltage reduces the average bit-line discharge energy and improves the Ion/Ioff ratio on the read bit-line.

SA with a limited Vt swing in the first stage : A sensitive SA allows the use of a very small signal on the bit-line, which reduces the cell read current that is required to achieve a given access speed and hence allows to improve the other cell metrics. The small swing also reduces the active energy consumption. The presented two-stage SA design combines a small offset voltage with low active energy consumption, which becomes more difficult as mismatch increases with further scaling. Although SA calibration can achieve similar improvements, it comes with additional test complexity. Additionally, the two-stage approach is more robust to other effects such as thermal noise and random telegraph noise (RTN), which put limits on what can be achieved with calibration.

Charge limited sequential sense amplifier: Large variations in drive current and timing due to process corners, temperature effects, and mismatch complicates accurate control of the swing in the first stage. Supplying the SA current from a pre-charged capacitor rather than from a normal supply provides a robust control over this swing.

Static write bit-lines: 8T cells enable the use of static write bit-lines, which reduces the average write energy without degrading any margins and without overhead. While low swing write schemes become less effective with scaling because of the increasing offset voltage of the local write receivers, static write schemes do not suffer from scaling, hence the advantage of static write bit-lines becomes more pronounced with scaling.

The techniques proposed address the most prominent issues associated with advanced technology nodes. They reduce the sensitivity of the design to **transistor mismatch:** a separate write port and lower read current requirement ease cell design, the automatic bit-line swing limitation avoids the need for an accurate duration of the word line pulse, and the new SA design makes sensitive SAs affordable from an energy point of view, even with larger transistor variations. The **active energy** consumption and the **leakage power** of the memory that implements the presented techniques are very low. **Electro-migration** becomes a significant issue for minimal width wires and single vias and contacts, which cannot be avoided within the SRAM matrix. Hence, the low write current associated with the HVT 6T core and the lower read current thanks to the sensitive SAs become advantageous.

8.3 Conclusion

Wireless sensor networks are transforming our interaction with the world. The embedded memories consume a major proportion of the power budget for the computation intensive wireless sensor nodes. The increased memory sizes allow computation intensive wireless sensor nodes to perform more complex signal processing and store more sensor data in order to reduce the data transmission. But today available embedded SRAM modules have a prime objective to be area and performance effective. With the result energy consumption is very high and is not suitable for energy limited wireless sensor nodes. This necessitates the SRAM design to be relooked. The design methodology for the SRAM modules for the energy limited wireless senor nodes prioritized energy efficiency and variability resilience over silicon area and clock speed.

The proposed circuit techniques have been validated with two silicon prototypes of embedded SRAM module. The first prototype (IM_90) of 128 kbit 6T SRAM in 90 nm LP CMOS consumes 4.4 pJ/access while operating at 80 MHz. The variability resilient and low power techniques featured in IM_90 include innovation in the local architecture with the use of local read/write assist circuitry. The energy efficient hierarchical bit-lines structure includes low swing global bit-lines and VDD/2 pre-charged short local bit-lines. The innovative MS-SA-R calibration technique for the global read sense amplifiers of the SRAM not only adds to the variability resilience but also yields maximum energy reduction compared with existing calibration techniques.

The second prototype (IM_65) further validates the effectiveness of the circuit design techniques in reducing the energy consumption of SRAM memories. The main design target is to reduce the energy consumption and to mitigate the impact of increasing variability for the advance sub-nanometric technology nodes. A 64 kbit embedded SRAM in 65 nm LP CMOS (IM_65) sets a record low energy consumption of 2.65 pJ/access at 90 MHz. The design innovations featured are. Reduced swing dual Vt 8T SRAM cell mitigates leakage by 7x. Write-ability is improved by the Mimicked Negative Bit-line technique, which reduces write failures by 10^3x. The energy consumption is further reduced by using a novel CLS sense amplifier, which achieves a σVoffset of 14 mV with energy consumption of only 11 fJ/decision, without requiring post-silicon tuning.

8.4 Future Directions

The proposed design techniques can be tried for the advance technology nodes viz. 32 nm, 22 nm, and 14 nm and can be further improved to cope up with the increasing variability with the technology scaling. A comparative analysis of proposed circuit techniques for their resilience against bias temperature instability (BTI) can be carried out as a part of future work. The impact of change in threshold voltage due to aging can be analyzed. The circuit design techniques for SRAM which should be resilient to aging effects can be investigated as a part of future work. Similarly SRAM circuit design with FINFET devices can also be explored. In FINFET devices short channel effects are suppressed by using thin body transistor structure, which facilitates gate length scaling down to 10 nm regime.

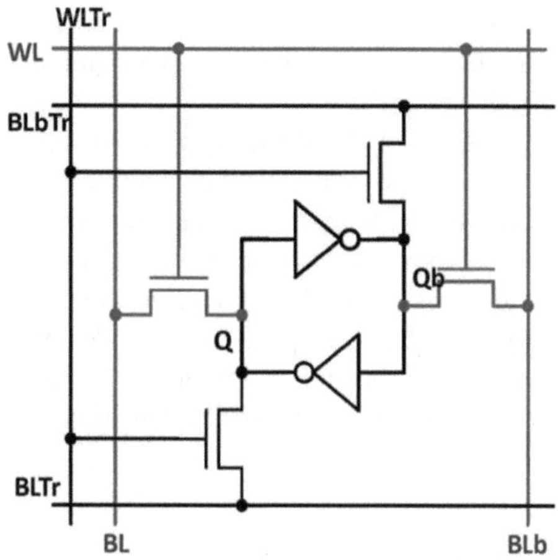

Fig. 8.2 Transpose SRAM cell (Seo et al. 2011)

In our prototypes area overhead is on higher side. The capacity/area for IM_90 is 0.09 Mb/mm^2 & 0.168 Mb/mm^2 for IM_65, which is 3–4 times less that the state-of-the-art high density SRAM modules. This area overhead can be reduced by utilizing litho optimized SRAM cells. Unfortunately, litho optimized SRAM cells were not available for this work. But for the future design exploration in order to reduce the area overhead the proposed circuit design techniques should be tried with the litho optimized SRAM cells.

At present, the SRAM prototype developed are full custom and are not compilable. Compilable low power SRAM design techniques should be explored. The parameters to be optimized for realizing compilable SRAM can be investigated. For example, the trimming of the word line drivers for the different word length configurations. The SRAM cell access transistors upsizing, for the different column heights etc. This will require reinvestigation of the low power circuit design techniques for SRAM from the compilation perspective.

The developed SRAM prototypes require more than one operating voltage IM_90 operates at VDD of 0.8 V and the local bit-lines at 0.4 V and 0.2 V for the highly capacitive global bit-lines. IM_65 utilizes, VDD = 0.8 V and 0.2 V pre-charge value for the bit-lines. The introduction of 0.2 V pre-charge value for the bit-lines result in a tremendous reduction in the energy consumption and an improvement in the variability resilience. The dual rail memories are becoming more acceptable for the advance technology nodes. But still the extra supply voltages result in the increased complexity of integration. The existing single supply SRAM designs (Kushida et al. 2009; Takeda et al. 2011) are very energy expensive. For the future design exploration single supply near Vt, low energy SRAM design techniques can also be explored. Alternatively, generation of multiple supplies with integrated DC–DC with SRAM can also be explored.

The SRAM circuits can be co-designed for the algorithmic and the architectural optimization for further reducing the energy consumption. For example, in a data parallel (SIMD, vector, sub word) approach in the processor data-path. The subwords are fetched and written back to the L1 SRAM in a parallel manner. The significant part of the total read and writes energy is "lost" in the periphery of small SRAM's.

The target SRAM design has to support the algorithmic changes done to solve the above problem for data parallel access. The conventional SRAM arrays are accessed only in rows. The column-based access would require an inefficient, energy expensive serial operation. To solve this problem, a single-cycle write and read access in both row and column directions (transposable SRAM, Fig. 8.2) is required. The transpose memory design should be explored to meet the ultra low energy access and to efficiently support the algorithmic modifications for data parallel access of 2D (or higher dimension array's).

References

K. Kushida et al., A 0.7V single-supply SRAM with 0.495um^2 cell in 65nm technology utilizing self-write-back sense amplifier and cascaded bit line scheme. *IEEE J. Solid-State Circuits* **44**(4), 1192–1198 (2009)

J. Seo et al., A 45 nm CMOS Neuromorphic chip with a scalable architecture for learning in networks for spiking neurons, in *Proceedings of IEEE Custom Integrated Circuits Conference (CICC)*, (2011)

K. Takeda et al., Multi-step word-line control technology in hierarchical cell architecture for scaled-down high-density SRAMs. *IEEE J. Solid-State Circuits* **46**(4), 806–814 (2011)

MIX
Papier aus verantwortungsvollen Quellen
Paper from responsible sources
FSC® C105338

If you have any concerns about our products,
you can contact us on
ProductSafety@springernature.com

In case Publisher is established outside the EU,
the EU authorized representative is:
Springer Nature Customer Service Center GmbH
Europaplatz 3, 69115 Heidelberg, Germany

Printed by Libri Plureos GmbH
in Hamburg, Germany